VALLEY EMPIRES

Hugh Glenn and Henry Miller in the Shaping of California

Ann Foley Scheuring

Library of Congress
Control Number 2010900162

Scheuring, Ann Foley
Valley Empires: Hugh Glenn and Henry Miller
in the Shaping of California

Includes illustrations, bibliography, and index.

ISBN 978-0-615-34123-1

Published by
Gold Oak Press
PO Box 7
Rumsey, California 95679

Book and cover designs by Jay Leek
Cover photo courtesy of Al Hayes
Printed in the USA

History is biography
writ large.

Contents

List of Illustratons

Preface and Acknowledgments

CALIFORNIA BURST ON THE WORLD SCENE with the Gold Rush of 1849, attracting ambitious men from all over the world, but not all of them were interested in mining. Two of the newcomers, Hugh Glenn and Henry Miller, were extraordinarily enterprising individuals who left deep imprints on state history because they focused their energies on agriculture and became profoundly involved in shaping the economic and physical development of the great Central Valley. They were very different men—one from a well-connected family and trained as a physician, the other a penniless and poorly educated German immigrant—but both were blessed with remarkable abilities and became famous for their accomplishments in the nineteenth century.

My interest in their stories was aroused partly because neither Glenn nor Miller seems well known today, and partly because my family had a personal link to their activities in Oregon. In Harney County, Oregon, where I was born, my grandfather worked as a ranch superintendent for Henry Miller's Pacific Live Stock Company in the 1890s before he began ranching for himself in that high desert country, and like others he knew Miller's habits of close management. One of my grandfather's friends also served on the jury that tried the

Peter French murder case, which effectively ended the dominance of the great P Ranch established by Hugh Glenn also in that country. Despite these Oregon connections, Miller and Glenn were primarily California ranchers and businessmen–and much more, as I came to find out in exploring sources to satisfy my curiosity.

Today the huge holdings of Hugh Glenn and Henry Miller are legends of the past, for the twentieth-century construction of massive public water distribution systems has greatly altered the early patterns of California agriculture. The stories of these two titans are well worth revisiting, however, for against the backdrop of struggling times their characters demonstrate both the nature of leadership and the pitfalls of ambition.

In recounting the results of my journey into the past, I wish to acknowledge the help of the always accommodating reference librarians at the Bancroft Library at UC Berkeley; the Department of Special Collections in Shields Library at UC Davis; the California History Room of the State Library in Sacramento; the North Baker Research Library of the California Historical Society; the San Francisco Public Library History Center; the Huntington Library; and local archivists and history lovers in several communities, including Charles Sawyer of the Milliken Musem in Los Banos; Evelyn Whisman of the Willows Museum; Tom Howard of the Gilroy Museum; and Mel Russell of the Yolo County Archives. Readers of all or parts of the manuscript who gave valuable feedback but are in no way responsible for my errors include John Chrisman, Dennis Dingemans, Marcia Kreith, Chris Scheuring, Charles Tays, and Rene Viargues. My most heartfelt gratitude goes to my husband David Scheuring, fountain of first-hand information about the challenges and joys of farming, and willing to drive long distances in pursuit of resonant landscapes and stray pieces of information. He has given years of patient support to this long-drawn-out project.

Introduction

THE HEARTLAND OF CALIFORNIA is its great Central Valley, that immense bathtub-shaped alluvial depression that lies between the coastal mountains bordering the Pacific Ocean to the west and the lofty Sierra Nevada separating the state from Nevada to the east. The valley is about 450 miles long and an average 50 miles wide, comprising about 15 million acres—the largest geographically distinct and most productive valley on the continent.

The Central Valley consists of two interconnected river basins known as the Sacramento and San Joaquin valleys. The Sacramento River runs from north to south, the San Joaquin River from south to north, both streams meeting in the delta region that drains into the Suisun and San Francisco bays. The tributaries of both rivers flow mainly out of the Sierra. With the exception of Putah, Cache, and Stony creeks in the Sacramento Valley, no significant streams flow down from the Coast Range, and the west side of the Central Valley is noticeably drier than the east side. In the far southwestern corner of the San Joaquin, before the age of irrigation, the Sierra-fed Kern, Kings, and Kaweah rivers emptied into three huge seasonal lakes—Kern, Buena Vista, and Tulare—that once covered hundreds of square miles during wet seasons, but have since been converted into farmland.

Rich as the valley's natural resources have turned out to be for agriculture, from the beginning of state history their geographical distribution and the variability of the climate made settlement more difficult than along the coast. Some parts of the big valley were not developed until well after the mid-twentieth century brought massive public water transfer systems into operation. In 1850, at the time California became a state, the Central Valley was nearly all virgin land and very sparsely populated. In this great interior valley, full of wildlife, vast tracts lay open for decades, waiting for those who could effectively begin their conversion to farmland—and for the institutions that could help them do so. Unlike the more easily settled coastal areas, the big valley posed challenges that would require not only substantial personal resources but unified political will to resolve, forged through decades of negotiation and politicking.

The historical transformation of the great Central Valley has not attracted the kind of fascinated public attention that has followed development in San Francisco and Los Angeles, even though the Valley today is recognized as one of the world's most highly productive agricultural areas. Perennial controversies over water, farm labor, and modern agricultural technology continue to agitate some public discussions, but relatively few of those who engage in such discussions are well versed in regional history.

This volume aims to open up some of that history. How did the Central Valley begin developing as an agricultural powerhouse? Who were some of the characters who shaped that development? What were their challenges, and how did they confront them? What can the past tell us about the present? Or even, perhaps, the future?

The hordes of immigrants to California during the Gold Rush included men who were less interested in mining than they were in meeting the steadily growing need for food. Two of these men, Hugh Glenn and Henry Miller, recognized the potential of the Central Valley even as its sweeping reaches still lay unbroken by fences or cultivation. Glenn and Miller would eventually help shape the development of the Golden State as profoundly as the mining kings, and their influence would last longer, for they would be the forerunners of the growers who have built today's farm economy—a multi-billion-dollar complex of industries that market food across the country and around the globe.

Hugh Glenn became the state's most prominent wheat grower as California entered the international grain trade, and Henry Miller's long career paralleled and in some ways surpassed those of the cattle kings of the Great Plains. Out in still-remote California, both men found ways to flourish through farming while their contemporaries built careers based on mining, finance, railroads, and petroleum. They were the rural counterparts of urban capitalists like Stanford, Huntington, and Hearst in California, and Vanderbilt, Carnegie, and Rockefeller in the East—but their careers were based on the productivity of renewable resources, the land and water that feed us all. They were able to overcome the challenges of fickle weather, limited water supplies, pests and disease, while they adapted to the uncertainties of variable markets, long-distance transportation, and changing labor supply.

In the Sacramento Valley Hugh Glenn was called the "Wheat King." A well-educated Missourian trained as a physician, he abandoned his profession to take an entrepreneurial chance on California during the Gold Rush, turning his energies at first toward the livestock trade. In 1867 he began growing grain on the Jacinto Ranch along the Sacramento River, soon accumulating more than 60,000 acres and building a national reputation. Glenn's career paralleled the wheat boom, and he even ran for Governor in 1879, though he lost the election. Four years later his murder abruptly ended his career.

In the San Joaquin Valley Henry Miller built a land-and-cattle empire during the same era. A German immigrant who began his career as a butcher's assistant, he quickly demonstrated the shrewdness and drive that enabled him to build a vast cattle ranching network with his partner Charles Lux. An extraordinary manager with a genius for detail, Miller eventually came to own more than a million acres in three western states and was the largest landholder in America at the time of his death. While Miller's empire became one of the legendary enterprises of the Far West, he also had a profound influence on western water law.

These two men pursued their ambitions in an era of great opportunity, and they helped shape the evolution of the Central Valley into a region of abundance. Both men made huge tracts of land productive, and both epitomized the ethos of nineteenth-century America, for they were on the cusp of industrial change in their scale of operations, adoption of new technology, and vertical integration from field to market. They exercised great influence over the heartland's history, and the ripples of that influence have spread far. As continuing environmental challenges put pressure on today's farmers and as portions of the Valley succumb to urban growth, our views of their legacy may differ; but their stories tell us that determination and commitment can mold local destinies in long-lasting and productive ways.

Prologue

Before the Gold Rush

ON THE FAR SIDE OF THE WESTERN WORLD, so remote and mysterious that the first reports resembled fables, the land called California entered modern history not much more than two short centuries ago. Separated by formidable mountains and deserts from the rest of North America and reachable by sea only after arduous voyages around the tip of South America, that part of the Pacific Coast lay almost inaccessible to early European explorers. They visited there scarcely at all during the great age of global discovery following Christopher Columbus. During nearly three centuries of European expansionism across the Americas, California remained dauntingly distant, thousands of miles beyond the continental frontiers being claimed by the British, the French, and the Spanish. Sixteenth-century Imperial Spain devoted its energy to conquering territory in South and Central America—which lay much closer to Europe than did California—and it was not until 250 years *after* Cortes had landed in Mexico in 1519 that Spain sent expeditions to colonize its northwestern frontier. Thus until the late eighteenth century the natural resources of California lay virtually undiscovered by outsiders.

Long before Europeans entered this far-off place, however, people were living on the land in what one historian has called "Malthusian equilibrium" with the environment.[1] Many groups of these early Californians had maintained themselves for centuries within distinct territories–in the aggregate, a larger population than existed in any comparably sized area in North America at the time. Comprised of small tribes or bands inhabiting areas circumscribed by particular ecosystems, for the most part they spoke different languages and practiced different customs. The diversity of Indian California developed largely because of the extraordinary diversity of the environment. Tribal enclaves ranged from the hot desert sands around the Colorado River in the south to the rain-washed coast of northern California, from the tule marshes and dry plains of the great interior valleys to the oak woodlands of the foothills and pine slopes of the mountains. Centuries of experience taught these native Californians to live in harmony with places that they knew intimately, in small communities organized around the availability of food, and they developed considerable skill in managing the resources that provided them with the basics for subsistence. Like hunter-gatherers elsewhere, however, they found themselves nearly defenseless against the coming of Europeans.

Juan Rodriguez Cabrillo, sent out to explore by the Viceroy of New Spain (Mexico) in 1542, made the first recorded sightings of the coast of Alta (Upper) California from the sea, making landings between San Diego and Point Conception but unable to find further anchorages. In 1579 the English explorer Sir Francis Drake touched at what is now called Drake's Bay, north of San Francisco, but his countrymen later aimed at land further up the Pacific Coast. In 1595 the Portuguese merchant-adventurer Sebastian

Rodriguez Cermeno explored part of the California coast on his way across the ocean from Manila to Mexico, and seven years later Sebastian Vizcaino identified Monterey Bay as a potential harbor for the Spanish government. But "[t]he lapse of 60 years between Cabrillo and Vizcaino, and of another 167 years between Vizcaino and actual Spanish settlement of California, emphasizes the utter isolation of California and the low regard in which the Spaniards held it." [2]

Not until the 1760s did Spain take steps to colonize Alta California, hoping to reserve the territory against other nations' incursions. In 1768 San Blas was founded on the western coast of Mexico to serve as the supply base for the land and sea forces setting out on these first efforts. Franciscan missionaries would be sent to convert and civilize the Indians, while the government would develop military presidios and civilian pueblos.

The overland approach through Sonora was delayed by Indian hostilities as well as harsh conditions, but beginning in January 1769 three ships sailed north from San Blas. Two arrived in April at a landing named San Diego; the third was lost. In mid-May the first overland party arrived–27 soldiers led by a Captain Rivera, driving a small herd of cattle, horses, and mules. The second overland party, led by Captain Gaspar de Portola, arrived at the end of June and included Father Junipero Serra, the father-president of the prospective missions. More than half of the 300 men sent north died, either along the way or after reaching San Diego. Father Serra devoted himself to caring for invalids in makeshift quarters, while Captain Portola pressed north toward Monterey Bay, taking most of those who could still travel. His party passed through the sites of the future Los Angeles and Monterey, and in October reached the as-yet-unknown bay of San Francisco, sighting that great body of water from the mountains of the peninsula enclosing it. Portola's company camped here, trying to decipher their surroundings, but in November gave up explorations to limp back to San Diego.

Five years later, in 1774, Captain Juan Bautista de Anza finally opened the land route into California from Sonora, struggling across the Great Colorado Desert between the mountains of northern Mexico and those of the coast, but laying out a usable path to Monterey Bay and, eventually, to San Francisco Bay in 1776. Anza's second expedition brought in about 300 head of cattle to supply the new missions. Nevertheless, Indians massacred some two dozen Spaniards near the Colorado River in 1781, and for years thereafter travel between mainland Mexico and inland California remained minimal. Spanish settlement took place within a relatively narrow strip along the coast, with settlers and supplies brought in largely by sea.

Although a few Spanish explorers entered the San Joaquin portion of the Central Valley and reported seeing rivers and marshland, their forays in the great interior did not extend as far north as the Delta. Their mapping was inaccurate, and their mainly unfavorable impressions were based on the seasons in which they travelled. Not until Gabriel Moraga led expeditions eastward from the coast between 1805 and 1817 did the Spanish penetrate the Central Valley with the purpose of evaluating its resources for mission sites. These explorations identified a few possible sites in the eastern San Joaquin Valley, though not in its western or southern reaches, but concluded that none seemed promising enough to be worth the effort it would take to develop them.[3]

Over a period of 30 years the extraordinary Fra Junipero Serra built a chain of missions along what came to be called El Camino Real, the "royal road" north of San Diego, while the Spanish government established four presidios with garrisons and three pueblos. Among these widely separated settlements, however, lived only about 100 colonist families. A few military veterans received land grants from the Crown, but not many Spaniards found the prospect of living in colonial California appealing. For half a century the province remained an extreme frontier of imperial Spain, in little contact with the outside world, its resources largely unexplored.

The mission system defined Spanish California. Franciscan friars came to the New World to spread the teachings of Jesus Christ, but they also introduced the culture of Western Europe. The missionaries were able and energetic men, often gifted in very practical ways, and they undertook some noteworthy enterprises to support their missions, which had to be self-sustaining and prepared to feed, clothe, and house hundreds of converts as well as the handful of supervising Spanish. The colonial government allocated cattle to the missions as well as to the presidios and individual settlers, but it was the Franciscans who introduced Spanish-style agriculture into California. Though limited by local conditions, the friars built irrigation systems and raised poultry, sheep, and hogs as well as crops of wheat, corn, and beans. They also brought in grapes, olives, walnuts, and figs, which adapted well in California. Even with crude farming methods, the missions soon produced enough food and fiber not only to sustain themselves but to trade. With control of a docile labor force, some missions eventually grew wealthy.

Believing themselves sent by God to save souls, the missionaries focused on converting the Indians, who responded to kindness or were intimidated by threats, and who must have been impressed by the priests' knowledge and energy. These simple souls became neophytes bound to service, trained to farm, garden, and care for livestock, and to practice trades and crafts like carpentry and weaving. As some tribes converted to white men's ways, their old ways of life began to fade. Hundreds of adults and children became dependent on each mission. At its peak in the 1820s the California mission system extended from San Diego to Sonoma, encompassing a total of 21 missions with an associated population of more than 30,000 Indians—small but productive outposts of Western European civilization.

Spain was still clinging to its imperial dreams as the nineteenth century dawned, but the young United States was a rapidly developing nation, pushing into the sparsely populated regions of North America beyond the Appalachians and hungry for the land beyond its formal borders. Though the continent was yet unmeasured, American leaders knew that Europeans harbored ambitions for the western territories. The Southwest was already in the hands of the Spanish, British fur traders were making inroads on the Far Northwest, and Russian ships had established a port in Alaska. In 1803 the huge Louisiana territory fell fortuitously into American hands when Napoleon sold millions of acres to the U.S. for a pittance. President Jefferson sent the Lewis and Clark expedition out of St. Louis, then a muddy outpost on the Mississippi River, to map the lands acquired by the purchase and to try to find a route to the Pacific Ocean. Lewis and Clark's Corps of Discovery charted a huge swath of the continent but never dropped south of the mouth of the Missouri River.

About this time, however, a few foreign vessels were appearing sporadically along the coast of California. In the 1790s British Commander George Vancouver stopped in California several times on his surveys of the coast, but he did not admire what he saw of Spanish settlement and concentrated his efforts to the north where the country still lay open to the British. In 1812 Russian seal hunters established a tiny though never very successful coastal station at Fort Ross, 60 miles north of San Francisco. New England whaling ships also began to test California waters.

Small wooden sailing ships, however, were ill equipped to challenge what usually seemed an inhospitable coast, and those who landed were not in general favorably impressed. Otto von Kotzebue, a Russian mariner who visited San Francisco Bay in 1816, reported: "The environs of San Francisco, in the northern hemisphere, are much poorer in natural productions than the coast of Chili, under the same latitude, in the southern. In the spring, when winter has afforded the earth some moisture, the hills and valleys are indeed adorned with brilliant iris and other flowers; but the drought soon destroys them....The fogs, which the prevailing sea-winds blow over the coast, dissolve in summer over a heated and parched soil, and the country exhibits in autumn only the prospect of bare scorched tracts, alternating with poor stunted bushes, and in places, with dazzling wastes of drift sand." [4]

Such observations by passing seamen were not likely to attract visitors from afar, nor was the fact that the rolling fogs and strong tides through the narrow entrance to San Francisco Bay brought many ships to grief. Until the 1820s Spain's far frontier had little maritime commerce, for from the seaward side much of the coast seemed bleak. From the landward side, however, California also seemed inaccessible and drew scant attention from Americans for years, for it lay far beyond the enormous natural barriers of the Great Plains, the Rocky Mountains, and the western deserts. These continental hurdles continued to discourage overland explorations, while conflict with resistant Indian tribes also blocked general travel.

In 1821 Mexico won her independence from Spain in a fairly bloodless revolution, but for nearly a generation little changed in California, and some conditions grew worse. The Mexican government, stretched very thin, was largely ineffective, though jealous of its prerogatives, and life under its rule was even less progressive than under that of Spain. Some mission Indians grew restive under their enforced peonage, escaping from the control of the priests. One Miwok Indian leader of several uprisings in the San Joaquin Valley, called Estanislao, became adept at guerilla warfare and was able to discourage inland settlement for a time.

American fur traders trying to compete with the British were the first after Lewis and Clark to penetrate the immense country lying between the Great Plains and the Pacific. Some of the mountain men trapping along the Rocky Mountains eventually worked their way into California. The first was Jedediah Strong Smith, who made an amazing appearance at San Gabriel Mission in November 1826 after an exhausting trek across the uncharted high desert between Salt Lake and California. Smith was an explorer as well as a fur trapper, and his hardiness and courage were legendary. His skills allowed him to traverse portions of the continent that no white man had seen before. Before he was 30 he managed to identify South Pass as the preferred route over the continental divide and to explore more of California and the Pacific Coast than any American before him. His was the first recorded crossing of the Sierra, the first journey through the San Joaquin Valley, the first trek across the northern California mountains into Oregon—without maps, with minimal supplies, and with few men to back him. Smith escaped death a dozen times before being killed in 1831 by hostile Comanches west of Santa Fe.[5]

Smith was also the first to report favorably on the resources of the Sacramento Valley and thus to open that area to British and American notice. Following Smith's visit to Fort Vancouver after his 1828 trip through California and Oregon, Hudson's Bay Company trapping parties moved south to check out his reports of abundant wildlife. Between 1830 and the early 1840s British brigades of 60 to 100 men hunted and trapped through Oregon and northern

California. Impressed by reports of beaver, elk, antelope, and bear, more American trappers moved into California. One of them, Ewing Young, in 1837 drove several hundred head of Mexican cattle north from San Francisco Bay to provide Willamette Valley settlers with livestock.

The passage of trappers through the Sacramento Valley introduced "intermittent fever" or malaria into California during the 1830s, apparently through the vector of the marshland mosquitoes. Its impact was devastating. By 1840 thousands of Indians had died of the disease. Survivors were left with even less ability to withstand foreign intruders.

Meanwhile, the Mexican government, increasingly jealous of the power and wealth of the missions, issued a decree in 1833 that began the secularization and granting of mission lands to private owners. These government land grants were a kind of "grand Mexican homestead law," according to an early historian.[6] Provincial governors hoped that more Mexican settlers might circumvent the movement of other nationals into the region. Most of the grants followed the chain of missions up the coast, with few at first in the inland valleys or north of Sonoma. Grants of up to eleven square leagues (a square league being equal to about 4,438 acres) could be obtained by petition at no cost if the petitioner would simply agree to occupy the land. Though the grants were never actually surveyed under Mexican rule, title papers were duly signed and recorded, based on hand-drawn maps called disuenos outlining vague boundaries. At the time no particular precision seemed needed, since the land was considered of relatively little value in its natural state, serving mainly as grazing ground for the half-wild cattle that dotted the countryside. And there was far more land to grant than petitioners who wanted it.

Most of the mission lands became private ranchos, which grew in number from about 30 in 1830 to some 700 by 1846 as the government tried to promote settlement. The transition from a well-organized mission system to private land ownership, however, sent the province into a general economic decline. Landowners continued to raise livestock and produce grain, often less efficiently, but the skills in handicrafts and trades developed at the missions were lost as Indian neophytes drifted away. As settlers appropriated the mission lands, stricken tribal groups retreated further into remote areas.[7] For property owners, however, the brief period of life after secularization was often quite pleasant. It was a pastoral time, and even under modest circumstances the Mexican ranchero was gracious and hospitable. Despite episodes of political conflict, the province generally drowsed in a kind of bucolic contentment, its people relatively indifferent to commerce. As one American noted, "Nature did everything, and man did nothing." These were the days of "cattle on a thousand hills," of genial life on broad ranchos for the fortunate few.[8]

The cattle of Mexican California were descended from the few hundred animals driven north by expeditions from Mexico. California's mild climate and broad grasslands allowed livestock to multiply rapidly, and by the 1820s as many as 100,000 cattle were grazing across the hills and valleys from San Diego to Sonoma, along with great numbers of horses and sheep. The rancheros depended on nature to provide grass and water for their herds and made little effort to improve their stock. While dry years could devastate the rangeland, most rancheros sustained themselves comfortably enough. Twice yearly they organized round-ups to bring the cattle in off the range and separate them out, brand the new calves, and mark some for slaughter. The Mexican *vaqueros* (loosely translated as cowboys) were highly skilled horsemen, and the picturesque round-ups were the highpoints of the year.

Cattle comprised the chief commodity in the economy for the entire Mexican era. Raised on open land with almost no expense, they were the primary resource to be bartered for other goods in the developing maritime trade. From 1822 on, the Boston firm Bryant and Sturgis regularly collected California cattle hides for the American market and, under strict regulation by the Mexican government, traded imported goods for both hides and tallow. This trade was mainly a kind of barter, in which the hides counted as "California bank notes" used as the medium of exchange, each hide being valued at about $1.50 in silver or $2.00 in trade. Although the Americans charged very high prices for the commodities and manufactured goods they brought in, the Mexican buyers snapped up what was available, paying as much as 300 percent over cost for clothing, shoes and boots, saddles, sugar, spices, tea and coffee, crockery and other items.

A seaman on a Boston trading ship, Richard Henry Dana, spent a year along the California coast in 1835–36 and later described the experience in his book *Two Years Before the Mast*, published in 1840 and an immediate best-seller. According to Dana (expressing a typically Yankee view), life in the province was both charming and inefficient: "The Californians are an idle, thriftless people, and can make nothing for themselves. The country abounds in grapes, yet they buy bad wine made in Boston and brought round by us, at an immense price...Their hides too, which they value at two dollars in money, they give for something which costs seventy-five cents in Boston; and buy shoes (as like as not, made of their own hides, which have been carried twice round Cape Horn) at three and four dollars...In the hands of an enterprising people, what a country this might be!.... The Americans...and Englishmen, who are fast filling up the principal towns, and getting the trade into their hands, are indeed more industrious and effective than the Spaniards..."[9]

California may have seemed a backwater in the 1830s, but America was changing fast. The national population was growing rapidly, swelled by waves of immigration from Northern Europe, and Americans were eagerly pressing westward, drawn by opportunities for land ownership and business in a country that seemed limitless. Savvy entrepreneurs could sometimes make fortunes, fueling the ambition of others. New steamboats and railroads helped many satisfy their desire to push west in search of better lives. As overland trade grew along the Old Spanish Trail between New Mexico and Los Angeles, some of the restless began to focus on the potential of distant California, encouraged by the few reports that came east. Though it was still a somnolent land governed from Mexico, dozens of Americans decided to migrate there to start farms and businesses. Some became Mexican citizens and married into California families.

Abel Stearns, already a naturalized citizen of Mexico, arrived in Los Angeles in 1829 to build a career as a prosperous trader and pueblo official. When he married the beautiful Maria Bandini, he acquired substantial property and soon became one of the wealthiest ranchers in the province. William Wolfskill, an American mountain man, entered southern California in 1831, also married a Mexican woman, and developed a well-known ranch near Los Angeles, experimenting with crops including grapes and citrus. He is credited with beginning California's citrus industry by planting orange trees from the San Gabriel Mission in 1841. To the north, American trader Thomas Larkin settled at Monterey in 1832, where he became a successful merchant and official representative of the U.S. government. His children, born in California, received one of the first land grants in the Sacramento Valley. Other Americans began to take up land beyond San Francisco Bay. In 1836 George Yount received a land grant in the Napa Valley, where he farmed and operated grist and saw mills, and John Marsh established a ranch near Mt. Diablo in 1837. These Americans brought a new viewpoint to bear on the development of California. Marsh even began promoting more American immigration through letters sent to eastern newspapers.

The most important settler in northern California during the Mexican era was John Sutter, an ambitious Swiss immigrant who laid the foundations for an empire in the lower Sacramento Valley in the 1840s but achieved only fleeting success. Born in 1803, Sutter faced financial failure as a young man and set sail for America in 1834, leaving his family behind. His next few years were a series of adventures as he pursued trading schemes in Missouri, Santa Fe, and the Oregon territory. He even spent four months in Hawaii, where he persuaded some backers to finance him in an effort to found a colony in California. In August 1839 Sutter and a crew of eight Kanaka workers explored their way up the Sacramento River, finally landing at a spot along the American River just past its confluence with the Sacramento. There they laid out a settlement at a place favorable for development.

Sutter had great ambitions but few resources of his own to realize them. He polished his gift for persuading others to back him or join him. His affability and determination were such that after only one year in California the 37-year-old Sutter was able to convince the Mexican government to appoint him as the commander of an official fort at his settlement, which would serve as the northern outpost for the province. The next year Governor Alvarado granted Sutter a domain of 11 leagues, approximately 48,000 acres, its borders defined by the Sacramento, American, and Feather rivers. Sutter named his colony New Helvetia with the hope of attracting Swiss settlers, and arranged for twelve families to settle there. It quickly became a center for agricultural development, with livestock and grain crops tended through Kanaka and local Indian labor. To provide supplies for the settlement he purchased Fort Ross from the Russians, who now wished to abandon it.

Sutter gathered resources on credit and went deeply into debt, but his energy produced rapid results. One historian writes that he "...created an establishment that was at once a well ordered military post, an active industrial center, and the seat of the most ambitious and varied agricultural development in California....he planted large areas to wheat; built a flour mill; diverted water from the American River for irrigation purposes; grazed large herds of cattle and horses, sent hunters into the mountains and along the rivers for furs and elk skins; set up a distillery; began the weaving of coarse woolen blankets; ran a launch regularly for freight and passengers between his settlement and San Francisco Bay...in short, made his colony the nucleus of all activity...in what was then the only settled portion of interior California." [10]

Following the opening of the overland trail into California in the 1840s, Sutter's Fort became a destination for American immigrants. In 1841 the first overland company to reach California from Missouri was the Bidwell-Bartleson party. Despite severe privations this party completed the long journey over the Great Plains and the Rockies by taking a cutoff from the Oregon Trail, then struggled through the desert and western mountains until they were able to stumble into Sutter's Fort. Young John Bidwell, the surviving leader of that first California-bound party, had managed to pull the group through several crises even though they had little inkling of where they were going. During the next few years, as "Oregon fever" swelled the number of wagon trains moving west, more of those risking the journey would opt for the California cutoffs that began to multiply through the rugged country between the continental divide and the coast.

Not long after Bidwell arrived, Sutter hired him as his foreman and representative. As Sutter's loyal employee for eight years Bidwell learned much about the territory and met most of the influential players of that era in California. He spent more than a year dismantling Fort Ross and transferring its equipment to New Helvetia. In 1843 he made one of the first reliable maps of the Sacramento Valley after a long chase following horse thieves, and in July 1844 he took on the task of finding a land grant along the Sacramento River for the children of Thomas Larkin. Thus, despite the heat of mid-summer, the limited availability of water on the mid-valley plains, and the presence of Indian villages along the river, he was able to recognize the opportunities for settlement in the still largely unknown upper Central Valley. He applied for a land grant himself and helped locate land for other settlers at Sutter's behest.[11]

In 1844 Governor Manuel Micheltorena tried to stave off foreign immigration into northern California by approving 33 land grants in the Sacramento Valley, all of them located along waterways. One of those furthest north was the Jacinto Rancho of eight leagues, or about 35,487 acres, given to Jacinto Rodriguez along the Sacramento River south of Stony Creek. The Larkin Children's grant of 10 leagues, or 44,364 acres, just south of the Jacinto grant, helped secure much of the west bank of the river to private ownership. These grants would later figure prominently in the development of the Sacramento Valley.[12]

Overland migration to the Pacific Coast from the American interior increased steadily during the early 1840s, with most of the traffic moving along the Oregon Trail, part of which was officially surveyed under Lt. John C. Fremont heading a unit of the U.S. Army Corps of Topographical Engineers. Fremont's published account of his expeditions greatly encouraged westward migration and shaped his reputation as an explorer. Although the majority of the wagon trains leaving Missouri aimed for Oregon's Willamette Valley, the success of the Bidwell expedition encouraged others to set out for California. In March 1844 Fremont and his company, on their second exploratory survey, arrived at Sutter's Fort after a miserable winter crossing of the Sierra Nevada. The next year six overland parties arrived, bringing 180 Americans. Ever-hospitable John Sutter provided shelter and provisions as well as information and advice to the immigrants.

The first Americans to settle in the Sacramento Valley above the Feather River were William B. Ide and Granville Swift. Ide came into California in 1845 and began eking out a pioneer existence along the river north of what is now Red Bluff. Swift built an adobe structure about 1846 along the banks of Hambright Creek near what is now Orland, grazing livestock on the plains.

California's total non-Indian population increased from about 3,500 in 1822 to about 6,900 in 1846. This small number of residents included a growing proportion of Americans: by that year there were about 700 in the province, up from fewer than 20 in 1820, about 150 in 1830, and 400 in 1840.[13] Mexican officials recognized that incoming numbers of Americans might overwhelm the territory, but their own hold on California was so feeble that there seemed little they could do. Mexico had already lost Texas to the United States, and the newcomers were bringing a restless energy to the development of California's barely touched resources. With far-ranging ambitions, Americans were already beginning to shape the economic and social life of California. The Mexican government grew increasingly uneasy, with good reason.

"Manifest destiny"—a term first used by a journalist after the annexation of Texas in 1845—became the popular catchword for the nation's aspirations to extend its dominion across North America.

Though other nations already had claims on some of that territory, Americans became convinced of the logic of their own continental expansion. Political advocates of additional annexations—of New Mexico, California, Oregon, even Cuba—persuaded themselves that it was the design of Providence that the United States should include all the territory within its scope. California lay squarely within that design.

As early as 1837 President Andrew Jackson had made an unsuccessful bid to purchase a portion of California from Mexico, but it was not until after the inauguration of James Polk as President in 1845 that the territory came into American hands. From the beginning of his term Polk wanted to acquire California, which was rapidly losing all semblance of competent governance from Mexico City, but he hoped first to be able to buy it, or, alternatively, to encourage Californians to declare independence and seek the protection of the American government, as in Texas. His hopes were stymied by months of unsuccessful U.S. diplomatic maneuvers and Mexican civil discord. Frustrated, Polk initiated a conflict in which America's neighbor was literally forced to cede territory. Congress declared war on Mexico in May 1846 on the pretext of a boundary dispute. According to one historian, "...Polk baited Mexico into war over the Texas boundary question in order to get California, after concluding that Mexico would not sell California." [14] In March 1847 General Winfield Scott's army landed at Vera Cruz to begin marching toward Mexico City, taking it over in September.

Before official news of the war had even reached California, however, a brief local insurrection took place known as the Bear Flag Revolt. In June 1846 a few Americans from the Sacramento Valley rode to Sonoma to take retired General Mariano Vallejo under arrest and declare California an independent republic, hoisting a crudely improvised flag over the plaza. John Fremont and his armed men appeared at the scene, Colonel Castro sent soldiers to reestablish order, and several men were killed in the skirmish. Fremont declared victory and assumed military command over northern California, then commandeered the disorganized presidio at San Francisco and sent General Vallejo into jail at Sutter's Fort. The "Bear Flag Republic," however, lasted less than a month. Commodore Sloat with the official U.S. Pacific naval squadron seized Monterey and raised the American flag there on July 7, then moved on to San Francisco and finally to Sutter's Fort, where the flag went up on July 11 and General Vallejo was released.[15]

The California phase of the war with Mexico ended after just six months. During the course of the conflict General Stephen Kearny marched troops from New Mexico to California; brief hostilities in southern California brought deaths both Mexican and American; and in January 1847 at Cahuenga, north of Los Angeles, Commander Andres Pico surrendered to Fremont and his California Battalion of volunteers.

The war with Mexico was formally concluded on February 2, 1848, with the Treaty of Guadalupe Hidalgo. Mexico ceded New Mexico and most of Arizona, Utah, and California to the United States for the trifling payment of some $15 million. The treaty was signed two weeks after gold had been discovered at Sutter's mill on the American River, but no one knew that then.

---

1 Rawls and Bean, *California: An Interpretive History,* p. 5. Sources on Indian California include Powers, *Tribes of California;* Cook, *The Population of the California Indians*; and Anderson, *Tending the Wild.*

2 Rawls and Bean, p. 17ff.

3 Eigenheer, *Early Perceptions of the Agricultural Resources of the Central Valley,* cites the generally negative early Spanish impressions of California's interior.

4 Descriptions of the early San Francisco area by Vancouver and Kotzebue are included in the selections compiled in *San Francisco Yerba Buena: From the Beginning to the Gold Rush, 1769–1849.*

5 See Morgan, *Jedediah Smith and the Opening of the West.*

6 Hittell, *The Resources of California* , pp. 453–61.

7 The total Indian population of California on the eve of the Gold Rush has been estimated by Cook at about 150,000, down by half from the early Spanish period.

8 Quotes from Dana, *Two Years Before the Mast*; Cleland, *Cattle on A Thousand Hills.* Cleland gives an account of the Mexican rancho period; Dana is the classic source on pre-Gold Rush California seen through American eyes.

9 Dana, pp. 86, 188.

10 Cleland, *March of Industry,* p. 22. Cleland's account is more flattering than that in Albert Hurtado's *John Sutter.*

11 See Gillis and Magliara, *John Bidwell and California*, for the life of this respected early settler.

12 McGowan, *History of the Sacramento Valley,* extensively covers the development of the valley.

13 Population estimates from Cowan, *Ranchos of California*, p. 10; McGowan, ibid.

14 Morison, *Oxford History of the American People*, p. 561.

15 Josiah Royce gives a detailed account of the Bear Flag revolt in his history of California written 30 years later.

Illustration details from "Crossing the Plains,"
a print depicting landmarks and phases of travel on the emigrant trail.
Courtesy of the Center for Sacramento History, McClatchy Collection.

~1~

Coming to California

ON JANUARY 24, 1848, a master carpenter named James W. Marshall noticed some color in the millstream of the lumber mill he was building on the American River for John Sutter. Gathering some of the small, shiny mineral flakes in the water, Marshall shared the discovery with his employer after a rainy 40-mile journey down to Sutter's Fort. Sutter performed a few tests after consulting his office *Encyclopedia Americana* and determined that the sample was high quality gold. It was an amazing find, and the two men hardly knew what to do about it. They tried to keep it secret long enough to finish the work they had underway, but the word leaked out. Within a matter of days many of the laborers at the Fort abandoned their jobs to hunt for gold in the water near the mill site. Within a month dozens of excited men were swarming along the gravel bars and river banks, rinsing for "pay dirt" in simple metal pans.

The discovery soon became an item of gossip in the little port of San Francisco. The tireless Mormon businessman Sam Brannan, who at the time was operating a tiny store near Sutter's Fort, is said to have marched through sleepy San Francisco holding up a jar of gold dust and shouting "Gold!" as a way to buck up his trade. It worked. Eager speculators made their way up the Sacramento River, throwing together outfits for mining expeditions from Brannan's newly enlarged store and heading for the hills. The first public announcement of the finding in a California newspaper appeared in mid-March. By autumn, as letters and dispatches came off ships from San Francisco returning to eastern ports, national newspapers began carrying sketchy reports. President Polk received a substantial sample of California gold by official carrier, and in his December 1848 speech to Congress remarked that this finding alone would repay many times over the cost of the war with Mexico. As the news flashed around the country, Americans were electrified.

This new development in the Far West now magnified its attractions for those who had already decided to venture beyond the continental divide, and beckoned seductively to those who entertained visions of quick wealth. During the next few years, hundreds of thousands of newcomers would flood into California. Full of energy and greedy for riches, their arrival would change the territory forever.

Though California was now officially a part of the United States, it was still largely a blank place on the map for most Americans, and the few reports available were immediately snapped up as guides. Early in 1849 many thousands of hopeful gold-seekers left home to make the difficult journey across the continent as soon as the weather permitted. Most of the '49ers were young single men on a quest for both wealth and adventure. Though some would find riches indeed, in the mines or elsewhere, many others would be sorely disappointed. Not a few would die in primitive circumstances on the trail or in camps after reaching California, their bodies buried far from home.

In 1849 two major overland trails led to the Pacific Coast, both departing from Independence, Missouri. The Santa Fe Trail followed the trail of early fur trappers and traders, heading southwest diagonally across the Great Plains from Missouri into New Mexico, then, as the Old Spanish Trail, passing circuitously from Santa Fe up through the rugged San Juan mountains and northern Colorado Plateau before crossing more mountains to traverse the Great Basin and Mojave Desert into southern California. The Oregon Trail followed the route taken by pioneer missionaries and settlers bound for the Oregon territory in the late 1830s, heading northwest across the Great Plains and through the Rockies via South Pass, then continuing toward the Columbia River while several cutoffs split off toward California.

Since gold had been discovered in northern California, the Oregon Trail and its cutoff routes were preferred by most '49ers, though some wagon trains opted to go south. In both cases the inland continental weather allowed travelers only a limited window of months to get through the country before the killing winter set in. Beyond the problem of potentially harsh weather, the journeys were fraught with peril. With or without Indian hostilities along the way, the treks were nearly two thousand miles long and full of dangers ranging from deep ravines, high mountains, and punishing deserts to bad water and fierce wild animals. Even with good luck the journey could take from four to six months. Some of the country was grim and bleak, some of it astoundingly beautiful, but the travel from day to day was always a steady, slogging effort, much of it on foot. Successful crossing of the continent depended on intelligent preparation, competent leadership, wagon train cooperation, and personal strength and courage.

One of the thousands heading out to California via the Oregon Trail in 1849 was a young man from Missouri named Hugh James Glenn. He moved onto the trail with greater personal advantages than many of his fellow travelers. The "scion of one of Virginia's most aristocratic families," according to a later family description, Glenn was born September 18, 1824, in the Shenandoah Valley of Virginia, into a family that could trace several generations in America. His family line was said to be descended from the Normans who had invaded Scotland; his ancestors had moved to Northern Ireland and then with other Scotch-Irish emigrants to Chester County, Pennsylvania, in 1730. A few years later the family moved to Augusta County, Virginia.[1]

His parents, George Glenn and Mary Gleave Anderson, were relatively well off and indulgent with Hugh, their only child, a small, bright, high-spirited boy with red hair. When Hugh was seven years old, the family moved west again, this time to a settlement called Paris in north central Missouri, the seat of Monroe County about 100 miles northwest of St. Louis. This area, sometimes called the "Blue Grass Region" or "Little Dixie," had begun attracting settlers from Kentucky and Virginia in the early 1830s. Livestock farms prospered here on the rolling prairie and hills of the Salt River Valley. Hugh had a privileged upbringing for a boy in a pioneer community. His father became a successful businessman, and the family was well regarded. Educated in private schools, Hugh was able to graduate from high school and then worked briefly as a clerk in a dry goods store. About the time of his mother's death in 1845, he began to attend lectures at the recently founded McDowell's Medical College in St. Louis. He spent a year there preparing for a career as a doctor.

When the war with Mexico broke out in 1846, Hugh enlisted for a year's military service as a private in the Second Regiment of Missouri Mounted Volunteers. The troops moved out along the Santa Fe Trail to fight in New Mexico, giving young Glenn his first taste of the West beyond the Missouri River. He participated in several battles including those at Matamoros and Taos, but received no serious injuries and was honorably discharged in 1847. Returning to St. Louis, he resumed his medical studies for two more years and graduated with honors in the spring of 1849 with an M.D. from the Medical Department of Missouri University, by then a school of 154 students. In March that year, after an engagement of several years, Glenn married Nancy Harrison Abernathy, the daughter of a respected judge in Monroe County. He seems to have pushed the thought of practicing medicine aside at least temporarily, for only a few weeks after their wedding Glenn left his 22-year-old bride to join the Gold Rush to California. Borrowing money to buy a fourth interest in an ox team and provision himself for the months ahead on the trail, he hoped like all the other young men heading west to find enough gold to build a nest-egg for his future on his return.

April 1849 in Missouri was wet and cold, and newspapers reported that several thousand emigrants were waiting to move out onto the plains as soon as the land began to produce enough grass to feed the livestock pulling the wagons. In both Independence and St. Joseph, the main departure points along the Missouri River, hundreds of wagon trains camped for days before they felt ready to move, ranging from the many small (4 to 10 persons) and loosely organized outfits to the few very large (more than 800) and well equipped companies. Glenn and his partners formed one of the small units. A reporter for the *Saint Louis Republican* sent back descriptions of many of the companies to his newspaper, noting that dissensions and lack of sufficient planning were preventing some groups from making a timely passage onto the trail. Those who departed late were threatened by an outbreak of cholera in Independence in early May, a disease that hovered over emigrants for weeks both in camp and on the trail.

Glenn's outfit started on the trail on May 3. Though little is known of their journey, the young doctor's medical and military training must have served him and his companions well. A pleasant-looking young man of medium height, he was fit and energetic, with a taste for adventure. Certainly it took grit and ability as well as weeks of bone-wearying effort to cross the continent to California. Along the way some travelers died of disease or injuries, their belongings scattered along the trail. Oxen and horses sometimes collapsed from exhaustion, and vultures and wolves lingered near the long lines of pack trains. Glenn's outfit made it through. In late August he and his partners arrived at Sutter's Fort, which had become the destination for many wagon trains as well as the main exit point for the gold mining areas.

Almost immediately Glenn headed for Murderers Bar on the north fork of the American River near Coloma, a chaotic place where hundreds of rough men crowded and quarreled over claims. He too staked a claim and put in two months of hard but unrewarding work panning for gold. Then, like many another enterprising man, he judged he could make more money by providing needed goods and services to miners than he could by competing in the frantic search for nuggets and flakes. He invested what he had left in an ox team to haul freight from the little settlement near Sutter's Fort to the mountain camps for a couple of months, then sold that business at a profit and established his own livery stable in Sacramento, the town that was rapidly being developed out of Sutter's land holdings near the river.

In October 1849 Sacramento had an estimated 2,000 inhabitants with a floating population of several thousand more. That fall the residents organized an election and approved a city charter. As winter descended, heavy rains began to flood the dirt streets leading up from the river, and hundreds of despondent immigrants, some sick, huddled in wet tents, easy prey for criminals and cheats. Boats docked along the river bank served as hotels, stores, and warehouses, while raw new wooden buildings were hastily put up for business on shore. Shallow-draft early steamboats plied the 120 miles between San Francisco and Sacramento in about six hours, bringing supplies for the grubby little town and the miners still in the hills. High freight rates for pack trains encouraged hundreds of teamsters, and soon eight-mule teams were hauling box wagons carrying four to six tons of merchandise along the rudimentary roads. Stage lines for passengers were not far behind.

High demand for good animals made livestock brokers and stable owners prosperous. The 25-year-old Glenn did a brisk business for a few months until he was able to sell his livery stable for what was in 1850 the goodly sum of $5000. With that money—which he later said made him feel richer than he ever did again—he purchased ship's passage out of San Francisco to New York and then, after buying gifts for his wife, returned to Missouri to reunite with his family. There he deposited his money in the Bank of St. Louis and became the father of his first child, a son. Having glimpsed the opportunities that lay in California, however, Hugh Glenn would not stay in Missouri long.

It is surely a matter of historical irony that John Sutter, the first settler in the Sacramento Valley, the colonizer and authority of the pre-American era, the genial gentleman and generous host to visitors and immigrants, was forced to sell much of his property at the end of 1849, his domain nearly destroyed by the chaos of the Gold Rush, the thievery and vandalism of gold-seeking newcomers, and his own ballooning debts. At the age of only 46, Sutter retreated upstream to his Hock Farm on the Feather River, while his newly arrived son helped sell off portions of his land grant to businessmen and investors in the emerging town of Sacramento. Many of the newcomers would make handsome profits, but Sutter, a dreamer and a drinker, declined into genteel poverty, his memories perhaps the greatest of his riches.

Unlike Sutter and most of the Mexican rancheros, several Americans who had settled in northern California before the influx of miners did well during the Gold Rush. John Bidwell, Sutter's long-time employee, worked another year for Sutter after the discovery of gold at Coloma, then left to search for gold himself along the Middle Fork of the Feather River at a place soon called Bidwell's Bar (now under Lake Oroville). After two seasons mining and running a store there, he was able to buy land and develop a large ranch in one of the most attractive areas of the upper Sacramento Valley, along Chico Creek. Although the appreciative Sutter offered Bidwell the hand of his daughter in marriage, Bidwell respectfully declined that honor, preferring to remain a bachelor until years later. At Chico he took the local village of Indians under his personal trust and began to engage actively in business and state politics.

Granville Swift, the livestock trader near Hambright Creek, also went to the Feather River, where he set a group of Stony Creek Indians to work separating gold from the river sands and according to Bidwell working them like slaves. By 1856 Swift stockpiled enough gold to leave the dry mid-valley plains for the more comfortable world of Sonoma County. William Ide, who had joined in the Bear Flag revolt and even been chosen as the president of the short-lived Republic of California, also acquired enough gold to buy property, but died of smallpox in 1852 before he could fully participate in the state's astonishing transformation.

Sacramento, a river town strategically located between San Francisco and the Sierra Nevada, served during the Gold Rush as the busy entrepot between boat traffic coming upstream and land transportation into the mining camps scattered along the western slopes of the mountains. Business opportunities during the early 1850s were excellent. As aspiring miners poured in, newly constructed streets teemed with wagon and pack trains carrying goods and supplies bound for the back country. The town quickly developed a large horse market where savvy men with good animals, like Hugh Glenn, could make handsome profits. Merchants like Sam Brannan did extremely well supplying necessities for the mining camps, and other Sacramento entrepreneurs amassed considerable wealth in a short period of time, quickly erecting substantial two- and even three-story buildings to serve as hotels, stores, and warehouses.

In the early 1850s four especially keen-witted young men, all in their 30s, began operating stores in Sacramento; eventually they would all become partners in a much bigger enterprise. Charles Crocker from Indiana first tried his hand at mining for two years and then established a Sacramento store selling dry goods and carpets. Collis Huntington from Oneonta, New York, who made money trading even while crossing the Isthmus of Panama, went into partnership with his friend Mark Hopkins in a highly successful wholesale hardware store. Leland Stanford, a handsome former attorney in Wisconsin who gave up law practice when his office burned down, came to California to look for gold but soon went into business with his five brothers. By the mid-1850s he too was operating a store and warehouse in Sacramento. Crocker, Huntington, Hopkins, and Stanford all prospered as the river town flourished, providing the growing population with needed goods while also building influential political connections in the new state capital as it developed there after 1854.

The frontier town had its full share of challenges. In October 1850 a cholera epidemic sickened hundreds and carried off an estimated 15 percent of the population. In November 1852 a large fire destroyed many of the town's new wood-frame buildings.

Throughout the 1850s periodic winter flooding from the two rivers bordering the community inundated much of the area developed between Sutter's original fort and the riverbanks, necessitating the building and rebuilding of levees and the regrading of streets time after time. For years Sacramento remained vulnerable to flooding along its waterfront.[2]

The southern overland route to California attracted some '49ers who eventually worked their way north. One of these was the remarkable Agoston Haraszthy, born in Hungary in 1812 to a family of the minor nobility. He was a well-educated and connected young man who had already toured the young United States and written a book about his travels before deciding for political reasons to emigrate from Europe with his family to settle in the wilds of pioneer Wisconsin in 1842. With substantial family wealth behind him, he founded a town on the banks of the Wisconsin River, which he called Szeptaj (Hungarian for Beautiful View) and others called Haraszthy Town (later renamed Sauk Center). Encouraging other immigrants to settle there, he started several businesses and experimented with farming in a brand new country. Despite his successes Haraszthy remained restless, and when the news of gold in California hit Wisconsin he persuaded his entire family, including his wife, six children, and his father, to head west. Always the entrepreneur, he never planned to engage in mining, a dirtier venture than he was attracted to, but hoped to seize on the opportunities that he knew would be plentiful as Americans swarmed onto the new frontier. The Haraszthy family took the Santa Fe Trail into New Mexico and then the Old Spanish Trail into California, arriving in San Diego at the end of 1849.

By all accounts a charming and persuasive man, Haraszthy filled a great void in San Diego, then a village of fewer than 700 people, mainly rough single men. He was personable, literate, full of energy and ideas, and he immediately made his mark by promoting real estate ventures, planting a fruit orchard and a vineyard, establishing a livery stable and a profitable butcher shop, and getting into local politics. In March 1850, in San Diego County's first election, he was elected sheriff. He was a tough officer: he cleaned up the waterfront, built a much-needed jail, and encouraged undesirables to leave town. The next year he was elected as San Diego's representative to the state legislature then meeting in Vallejo, where he succeeded in getting funds for improvements to the San Diego harbor and other projects. At the end of the session, however, having seen northern California, he sold his interests in San Diego and moved his family to San Francisco. The little port town was bursting with excitement, other Europeans were already living and working there, and he could envision great possibilities for the future. Haraszthy would soon be an influential player in the state's development.[3]

Many gold-seekers shipped out to California by sea. From ports on the Atlantic the difficult voyage around Cape Horn, the tip of South America, up to the port of San Francisco took as long as eight months. Though it was arduous, some came that way, but many more took a shortcut between the continents. The journey was thousands of miles shorter if taken through the Caribbean to the narrow Isthmus of Panama, where passengers could cross by land and then resume travel by ship northwestward along the Pacific coast. The Panama route was expensive, however, and not easy either. Besides battling seasickness in the Atlantic and making a strenuous 40-mile journey across rugged terrain by mule train or on foot, the travelers often had to wait weeks for ship connections on the Pacific side. During Panama's rainy season the water could pour down in torrents; footpaths turned into drainage channels, and the mosquitoes were fierce. Many travelers became ill with "intermittent fever" or "Panama fever," as malaria was called, and some died. Though the journey had its colorful aspects, it was usually with great relief that survivors moved on to their final destination in California.

San Francisco was the new name given to Yerba Buena in 1846. Richard Henry Dana had described the settlement of the mid-1830s as little more than a hide-shipping exchange, where sailors collected hides sent over from the missions and ranchos to trade for other commodities. Other early coastal travelers were also unimpressed with Yerba Buena under Mexican rule, though they recognized the enormous potential for shipping offered by the San Francisco Bay. "The whole harbor, besides its matchless qualities as a port of refuge on this surf-beaten coast, is the outlet of a vast breadth of fair and fertile land," Sir George Simpson, governor of the Hudson's Bay Company, had written on his way north in 1841,[4] and as other mariners gradually learned the idiosyncrasies of the coast and the bay entrance, they too began to appreciate the possibilities for commerce here.

By 1848, after the American takeover, the port of San Francisco—the hilly area surrounding the original Yerba Buena Cove—had become a small settlement with a few hundred inhabitants, but it was still a sleepy, bedraggled place. During the following year the little town almost literally exploded as the jumping-off point for hopeful gold miners and other visitors arriving by sea, ballooning to a population of an estimated 40,000 within two years. It was not so much a place to stay as a place to dock and depart again. In the jammed harbor dozens of vessels rode empty as passengers and crew members alike abandoned ship to set out for the mines of northern California.

"When you leave the ocean and enter the harbor of San Francisco it looks as if you were entering some noble river very much resembling the Hudson below West Point," wrote R.P. Effinger, a young Ohio lawyer, to his family in January 1850. "This neck or river is some six miles in length. After rounding a point, Yerba Buena and the great range of vessels now lying in this harbor are visible. The location of the city is on a high bluff....The whole harbor seemed to be alive with boats and yawls of every description.... When I landed at the long wharf (a kind of bridge running about six hundred yards in the sea), I was perfectly astonished at the immense crowd of people." [5]

Though much of the land beyond the harbor consisted of sand hills spotted with gnarled live oaks and hardy California lupine that bloomed only in spring, San Francisco rapidly became an "instant city," with rough cobbled streets, plank walks, and torch lights for night illumination. Merchants hoping to cash in on the traffic frantically assembled shops to attract customers. Around the harbor a motley collection of wooden structures, some not much more than shacks, rose to shelter businesses ranging from the legitimate to the unsavory. Like weeds after a surprise rain, hotels, saloons, and gambling houses sprang up to accommodate the newcomers. Within a couple of years Portsmouth Plaza, facing Yerba Buena Cove before it was filled in, was transformed from barren ground to a park with trees and flowers. Because lumber was scarce and some of the local carpenters were barely competent, a few forward-thinking newcomers arranged to bring in house kits or pre-fabricated sections of structures to assemble on shore. Some even imported corrugated iron doors and walls or quarried stone to defend interiors against the risk of fire, and more durable, ever-larger masonry buildings rapidly replaced the wood-frame constructions.

The section of San Francisco called the Gold Coast was named not only for the gold that came down into the city from the northern mines but for the large amounts of money brought in by travelers and spent in the sordid dives and gambling emporiums that took root along the waterfront. The rapidly growing town was a violent place. "Not a day goes by but what we hear of a suicide, a murder, or theft...Altercations are constantly occurring between parties about a gambling table, where out fly their Colt revolvers and an innocent man is just as likely to be killed as the guilty...men seem to live only for the gratification of their passions," wrote Effinger to his sister. San Francisco became such a center for criminal activity and corruption that reputable citizens soon formed vigilante committees to deal with trouble.

Lawlessness was not the only problem. Half a dozen major fires during the city's first two years damaged the settlement over and over again. Wind and highly flammable building materials, including the canvas tents pitched amid the crowded wooden shacks, exacerbated the flames that started in a hotel the day before Christmas 1849. The next year four more fires, some of them suspected arson, heavily set back rebuilding efforts, causing great losses. The sixth great fire, on May 4, 1851, also believed to have been started by an incendiary, was the worst. High winds fanned the flames to cause an estimated $12 million in damage to buildings, wharves, and even docked vessels in the water. These disasters impelled the young city to organize a fire department and establish policies to avoid such losses.

If the fires were not enough to unsettle the population, there were also earthquakes. Starting in 1849, earthquakes were recorded in San Francisco every few months, most of them mild but some causing severe damage to the new buildings and streets. Heavy shocks were felt in May 1851, shortly after the sixth great fire; again in November 1852; and in February 1856, when the water in the bay was reported to have risen and sunk perceptibly. Later major earthquakes in 1865 and 1868 also wrecked buildings and cracked the ground or caused filled areas to sink. Despite these unnerving tremblors, however, San Franciscans assured themselves that they were mostly something to laugh off and learn to live with, and relentlessly continued their pursuit of fortune.[6]

Hurrying through its application for statehood, California was formally accepted into the United States in September 1850. (Voters had already approved a state constitution in November 1849, though it would prove unsatisfactory some years later.) During its first ten years of statehood the population of California as a whole rose spectacularly, from an estimated 15,000 just before the Gold Rush to 93,000 in the 1850 census, then swelling to 380,000 by 1860. The newcomers hungered for wealth; some were well trained in trade and business, others were knowledgeable about livestock and farming, and their arrival marked a new beginning in the state's economic life.

More than 90 percent of the early population was male, and life was rough-edged as a result. But the state was also a remarkably cosmopolitan place from its beginnings, for it became a magnet not only for gold-seeking Americans but for energetic foreigners hoping for better lives. From Asia, South America, Europe, and the British isles they came. Many of the foreign-born brought unique skills from their homelands, contributing substantially to the rich mix of cultures and personalities that have marked both the state of California and the city of San Francisco from their earliest days. In the 1852 state census, the two national origins most reported were Irish and German, but as the years passed the city grew increasingly multi-ethnic and sophisticated.

The Germans who came to San Francisco in the 1850s ranked high among the early successful businessmen. They were part of the great nineteenth-century stream of German emigrants who sought both political freedom and economic opportunity in America. The first German immigrants arriving along the eastern seaboard in the

1820s were followed by thrifty German farmers seeking cheap land in the Midwest. By 1840 many Germans were established in trade and industry throughout the states. Between 1845 and 1860 more than a million more entered the country, outpacing all other immigrant groups, even Irish refugees from the Great Famine. The peak year for German immigration was 1851, when 221,253 arrived, many of them political refugees after the failure of the 1848 revolution. By 1860 half the foreign-born population of the U.S. was of German origin.

Germans flocked to California too. Some participated in the Gold Rush either as miners or as traders, while others set their sights on farming. Most were quickly accepted for their skills and work ethic. They were well organized too: the German immigrants of San Francisco's first decade established a benevolent society for mutual assistance as well as an insurance company, a savings and loan association, a hospital, a theater, a school, churches, musical groups, and restaurants. The Gesang Verein (choral association) that was organized in the city in October 1853 built a clubhouse at Post and Leavenworth that became an important meeting place. By 1860 two daily German newspapers were being published in San Francisco. The 1860 census, which listed about 70,000 residents, tallied 5,000 of German origin, the second largest ethnic group after the 12,000 Irish.[7]

Four German immigrants would eventually rank among the most successful of California's nineteenth-century businessmen: Henry Miller, Charles Lux, Isaac Friedlander, and Claus Spreckels. They would also help begin the astonishing development of California agriculture and play a part in the transformation of the great Central Valley.

The most remarkable success story was that of Henry Miller, the self-taught tradesman's son who would go down in California history as the "Cattle King," controlling an estimated million acres of land in three states and producing untold numbers of livestock for western markets. His was a rare convergence of opportunity and ability, and he made the most of it.

Born Heinrich Albert Kreiser on July 21, 1827, at Brackenheim, a small village near Stuttgart in the southern province of Wurtemberg, Germany, he was the youngest of four children and the only son of a butcher. Tradesmen prospered in this area, but his family members worked hard for their living, and their lives were narrow. Heinrich was born with an independent streak, and he grew up feeling no great loyalty for his homeland. Nor was he very fond of his exacting father, who forced his young son to work at distasteful chores from an early age. Heinrich was a homely boy—short-legged, squarely built, with a big head on a short neck, and small but piercing eyes—but he was highly intelligent, good at figures, and gifted with a prodigious memory. When he was only eight years old, his father set him to work as an apprentice in the meat trade, but he chafed under his father's harsh ways and had few friends. He spent only a few years in school, ending his formal education at the age of 13 to take on further responsibilities as his father's assistant. He learned a great deal in traveling from village to village to buy livestock for the family business, but he hated the idea of working seven years as a journeyman, the standard course of training to become a master butcher in Germany, and he yearned for more personal freedom. At the age of 15, after his mother's death, he left home for Holland. He spent three years working there, then another two in England. Along the way he learned to speak English, but never lost his thick accent or mastered all the grammar.

In 1847, like thousands of other young Germans, the twenty-year-old Kreiser booked passage on a ship to America, spending 60 days in steerage at sea (fifty of them sick, according to his later account). In New York City he found it difficult at first to break into the butcher's trade, which was controlled by gangs, so he spent three months working as a gardener for $4 a month plus room and board. He then found a job in one of the meat stalls in Washington Square, where he dressed out hogs and worked 16 hours a day for $8 a month. He was allowed to sell the offal independently to make more money, which he saved, and he slept in the back of the stall. Observing others making good profits, he determined to become an expert on everything pertaining to the meat business in America. After a few more months learning his way around, he found a job in an old man's butcher shop that was doing poorly and applied his energy to improving the business. His butchering skills and hard work added substantially to the profits of the proprietor, and when he took over the job of meat deliveries he became acquainted with a greater number of people and began to improve his limited English.

Heinrich Kreiser might have stayed in New York City to set up his own shop, but, according to Edward Treadwell, his first biographer, he was troubled by a developing romance and wanted to avoid an early marriage. Like others in the city, he and a friend named Henry Miller, a shoe salesman, were excited by the news of the Gold Rush in 1849 and talked of going there. In early 1850 his friend Miller even laid out $350 for a ticket for ship's passage from New York through Panama to San Francisco. Kreiser, who had been sending money to his family in Germany, did not have enough cash to join him, but when Miller suddenly decided not to go and offered him his ticket, Kreiser took it at a discount. On April 13 he left New York on the steamer *Georgia*. With the ticket issued in the name of Henry Miller and stamped "non-transferable," Kreiser got on board using that name, and he continued using it after arriving in California. A few years later he had his name changed legally from Kreiser to Miller.[8]

Miller's ship steamed through the Caribbean to land at the Isthmus of Panama during the height of the rainy season. The Panama rain forest was more exotic than anything the young German had ever seen, full of colorful bird life and monkeys swinging through the canopy, but travelers across the isthmus had to deal with both potential bandits and mosquito-borne disease, and, after they had passed over the mountains, found that they had to wait weeks in makeshift lodging along the bay for the limited number of vessels bound for San Francisco. To support themselves Miller and an acquaintance temporarily developed a small butchering business, but during the interval Miller himself became seriously ill with malaria and lay in a sickbed for more than three months. As he recovered, he found that his partner had not been able to carry on the business they had started, so he gave it up and worked his way, weak but determined, aboard ship up the coast to San Francisco, a 24-day journey. Later in life Miller remarked that had he not become ill he might have stayed in Panama, where he thought the opportunities to make money were very great.

Miller would remember that he entered the San Francisco harbor at 10 a.m. on September 24, 1850, with only his clothes, a walking stick, and six dollars in his pocket. At that time the port consisted of just four or five crowded streets, and much was blackened by the recent fires—the fourth such fire occurring only a week before Miller's arrival. Most of the resident population lived in cheap shanties or temporary tents upslope from the harbor, and the place was teeming with arrivals and departures. As soon as he limped off the boat, the young German began looking for work.

He quickly found a job as a butcher's assistant in a French shop on Jackson Street, shouldering heavy beef or pork quarters from delivery cart to shop in the crowded street and cutting up the meat for retail sale. Though still very thin and of only middling height, he was surprisingly strong, with remarkable powers of endurance, and behind his plain face was a keen mind, restless and aware. He had little interest in mining and knew he would be able to make money with his skills in a place where food prices were sky high. The town was awash with money, and gambling was one of the chief amusements, but Miller husbanded his earnings, again figuring out how to make a profit from the animal parts that the butcher would otherwise discard. Careful, conservative, and hard-working, he lived abstemiously, preparing his own food or eating an occasional meal of beef, potatoes, and coffee at a Chinese restaurant nearby.

While he was quick to take note of the opportunities in this wide-open place, others took note of his abilities. The meat trade was an essential business, and he knew the trade well. He was so fast and thorough at killing and dismembering animals—he later recalled that he once killed 100 hogs in a single long day—that one of his acquaintances offered to help him start his own shop. Not wishing to be obligated, he did not accept the offer, but instead began acting as a livestock broker, buying animals from the boats or down the peninsula and reselling them to retail butchers. He kept current on the meat market by keeping his ears open in the local saloons where the butchers and brokers gathered, participating in conversations but rarely drinking alcohol himself. Through shrewd dealing and with other backing, he eventually made a $33,000 investment in a herd on the peninsula—some of the first American-bred cattle in California—and with the profits from that investment soon afterward began his own enterprise in the city's business district. In 1854, at age 27, he rented a lot on Jackson Street between Dupont and Kearny and built a butcher shop and house, continuing to operate a small wholesale business on the side. He gained a reputation for selling good meat.

Charles Lux arrived in San Francisco in 1850 just a few months before Miller. Born in Alsace in 1823, Lux had come to America at age 15 and worked as a butcher's assistant in New York before becoming a U.S. citizen in 1848. In San Francisco, like Miller, he worked a year for another butcher and then went into business himself with his brother Henry, opening a retail shop at the junction of Kearny and Sacramento streets.

In 1853 Lux formed a short-lived partnership with a man named Edmondson to buy cattle wholesale and purchase a portion (1,700 acres) of the Buri Buri rancho in San Mateo County, a stunning piece of land between San Francisco Bay and the coastal range, where a tidal slough provided access to the property by water. The two were able to buy the property for $10,000, but three years later, as prices started to skyrocket, Lux bought out Edmondson's share for $18,000. He named his acquisition Baden after his birthplace in Germany. This site would figure prominently in the activities of the Miller & Lux partnership as it developed over time.

Isaac Friedlander was also making his mark in San Francisco in the 1850s. Born in 1823, this outsized character was a German Jew who arrived in New York as a boy and took the long way around Cape Horn to California in 1849. He tried his hand briefly at mining, but he made his reputation and his fortune as a grain trader during the wheat bonanza years, and eventually he also became a prominent investor in land and water development schemes in the Central Valley.

Friedlander was a towering figure—he was said to be 6 feet 7 inches in height—in more ways than one. By 1852 he turned from his first interest in gold to the marketing of flour and grain. Flour, like meat, had become scarce and expensive as California tried to absorb the enormous surge of new residents. During the wet winter of 1852 imported flour sold for as much as $50 a barrel wholesale in San Francisco, and even more at the mines. That year Friedlander strategically "cornered" California's imported flour market and made handsome profits for himself by forcing up the price of the flour. When small flour mills began springing up in scattered farming areas to compete with the imports brought in by sea, Friedlander invested some of his profits in several of those first mills and rapidly acquired a fortune. Then, seeing the demand, Friedlander built California's largest and most efficient flour processing plant of the time, the Eureka Flour Mills. Moving on to other ventures, he lost a considerable amount of money in speculation, but by 1858, the year that California's first grain shipments went around Cape Horn, he was back on the scene as an exporter of wheat going out of San Francisco Bay in chartered ships bound for both Britain and Australia.

By the mid-1850s both Claus Spreckels and Agoston Haraszthy were actively engaged in San Francisco business. Spreckels began his career in California in 1856 primarily as a merchant. Born to a farming family in northern Germany in 1829, he crossed the Atlantic at 19 years of age and worked for several years in the grocery business, first in Charleston, South Carolina, and then in New York City before coming to San Francisco. He and his two brothers, Diederich and Peter, began their business ventures in this city with two grocery stores, then invested in the Albany Brewery on South Everett Street between Third and Fourth. With its insatiable needs for alcohol, San Francisco by 1860 already had some two dozen small breweries, but the well-managed Spreckels plant soon outpaced the competition. Claus and his brothers operated the brewery only a few years before selling it for a considerable profit. Before long Claus would use his growing capital to begin a brand new industry for the West Coast: the production of sugar. Like Miller, Lux, and Friedlander, Spreckels too would leave a lasting mark on California agriculture.[9]

Agoston Haraszthy, who with his skill in languages must have moved easily in San Francisco's German circles as well as in general society, used some of his political connections to become the assayor at the U.S. Mint in San Francisco in 1855. During his three years in this position, a period of intense controversy—he was accused of embezzlement after mint records revealed missing gold, but exonerated when defective flues in the smelting process were found to be allowing gold dust to escape into the air—he began experimenting with the cultivation of wine grapes, first south of the city on the San Mateo peninsula and then in Sonoma County to the north. There, recognizing the potential for high-quality vineyards, he founded the Buena Vista Winery and continued his restless entrepreneurial activities while maintaining an office in San Francisco. Sometimes called the "father" of the California wine industry, Haraszthy would contribute greatly toward the development of the state's viticulture and winemaking.[10]

Men like these young Europeans, with drive, determination, and noteworthy talent, were ideally suited to seize the opportunities available in 1850s California. They were also lucky to be on the ground floor as the economy began to grow.

---

1 This narrative of Glenn's background is derived from several sources, including Allen, *Hugh Glenn*; White, *Glenn County Sketchbook;* James, "Dr. Hugh J. Glenn: The Forgotten Legacy"; scrapbooks and notebooks in the Willows Museum and Willows Public Library; and several Colusa County histories. Not all these sources are consistent in their details.

2 For history of the city of Sacramento, see McGowan, *History of the Sacramento Valley*, and Avella,. *Sacramento: Indomitable City.*

3 The definitive biography of Haraszthy is *Strong Wine,* Brian McGinty.

4 *San Francisco Yerba Buena,* p. 120.

5 From "R.P. Effinger's Excellent Adventure: The Unknown Letters of a Young Ohio Lawyer, 1849–1850," *California History*, 2004.

6 For early San Francisco history, see Lotchin, *San Francisco 1846–1856;* also Royce, *California.*

7 The San Francisco Main Public Library houses a collection of historical material relating to Germans in the city.

8 This narrative of Miller's early life derives from his interview with Edwin Fowler for Bancroft's *Chronicles of the Builders*; the interview is archived in the Bancroft Library and reprinted in *One Man Show.* Some details also come from Treadwell, *The Cattle King,* a fictionalized recollection of Miller's career and character.

9 Neither Friedlander nor Spreckels has received deserved scholarly attention, but Friedlander is mentioned in several publications about the wheat trade, including *Merchants of Grain*, and Spreckels is the subject of *The Sugar King in Hawaii* and *Claus Spreckels of California.*

10 See McGinty, *Strong Wine.*

~2~

Looking for Land

AN ESTIMATED TWO THIRDS of the state's burgeoning population was engaged in mining in the early 1850s. Many miners had originally been farmers, however, and once the placer mines began to decline about 1853, some turned their attention to the potential for farming instead. Certainly there seemed to be plenty of open land, an agreeable climate, and good prospects for developing markets. In the chaos of the rapidly growing new state, however, what seemed to be extraordinary opportunity to acquire land (at least for those with financial resources) was all too frequently stymied by one of the provisions of the Treaty of Guadalupe Hidalgo. Under the terms of the treaty the new government of California was required to recognize existing Mexican land titles so that landowners under the old regime might not lose their legal properties. As a result of this stipulation, pre-emptive land claims from the Mexican era complicated American attempts at settlement during much of the period through the end of the Civil War. Those who wanted to take up land for farming often discovered that the best pieces along the waterways were already blocked from entry. Mexican land grants in some areas lined the rivers a mile deep, and though much of this land appeared to be little used, it remained legally unavailable until the land claims were settled. "Squatting" on open land became fairly common, with many squatters hoping eventually to gain title to the land they moved onto.

One of the first actions the government undertook was to require that all land titles granted prior to California statehood be properly adjudicated. This was not easy. Most titles to huge land holdings–tens of thousands of acres in some cases–had been granted during the Mexican era on the basis of amateurishly drawn *disuenos* (maps) of the territory in question. Boundaries were uncertain, acreage was only estimated, and the claims themselves were sometimes questionably legal. In January 1852 the U.S. government established the California Land Grant Commission to validate claims, requiring claimants to present proof of title within a three-year period. U.S. attorneys were instructed to examine all petitions and require sound evidence of the legality of the purported grants. The burden of proof could involve lengthy documentation, court hearings with witnesses, and hefty attorneys' fees. Before a patent was granted, some claims might have to go through two or three court appeals, with years, even decades, of litigation and delay.

The patenting process, called a "vast injustice" to Mexican land grantees by contemporary observer John Hittell, sometimes proved too much for petitioners to bear financially. Portions of some grants were sold even while under adjudication, with further subsequent confusion. Landowners who had been well off during the Mexican era could become poor trying to prove their rights in the American era, and often found themselves disadvantaged in dealing with shrewd Yankees. Some went bankrupt and lost their patrimony altogether, while squatters on many of the broad ranchos damaged property without compensation to the owners.

Of the 813 land grant claims submitted by 1855 to the land commission, comprising a total close to 10 million acres, 440 grants were finally confirmed and 140 were rejected or abandoned by 1860, but 230 were still pending up through the late 1860s. Some claims were settled fairly early, others took many years. The Larkin Children's grant, 10 leagues near Princeton granted in 1844, was patented for 44,364 acres in December 1857. John Bidwell's claim for 22,214 acres near Chico, originally granted to William Dickey in 1844, was patented in April 1860. In the case of Henry Miller, who bought large portions, but not all, of the Las Animas grant near Gilroy, the ongoing litigation involved hundreds of owners and lasted nearly 30 years. The uncertainties were costly not only to the landowners but to the state. Hittell estimated that the delay in quieting land titles cost the state of California more than 200,000 permanent residents who might otherwise have settled here but were discouraged by the cloud hanging over potential land ownership for nearly a generation.[1]

U.S. government surveys provided the first extensive classification of Central Valley land in the 1850s but generally at first avoided evaluating any of the land grants in litigation. The uneven quality of assessments by the surveyors and the lag in sectionalization because of controversy over land-grant boundaries prevented these surveys from being useful to settlers for some years. Early surveyors tended to judge soil quality primarily by the existence of native vegetation, rating native oak woodlands along streams high in fertility but negatively viewing the treeless plains, especially on the west side of the Sacramento Valley and in many portions of the San Joaquin. Unwatered sections were considered of doubtful agricultural value, and marshlands were judged both useless and unhealthful—opinions shared by most newcomers during the Gold Rush years.[2]

Land titles and imperfect surveys were not the only challenge. Farmers used to climates with more reliable rainfall had much to learn about growing crops in California. The annual six to eight months of heat and drought in California valleys, with spotty availability of surface water, could challenge or defeat efforts at farming as they knew it, and building effective water storage and irrigation systems had to be a long-term, complicated, and expensive effort requiring social organization and financial capital as well as muscle and equipment. Other uncertainties included the lack of knowledge about the best crops and varieties to grow here. From the southern deserts to the coastal valleys, from the broad dry plains of the inland valleys to the rain-washed mountains of the northern coast, California was immensely varied, and what might grow well in one region might fail in another. The state's still small population offered only widely spread out markets for farm products, which had to be transported by wagon over rough roads or by boats operating on fickle rivers. Convenient transportation networks would require decades of large-scale investment by regional governments or private companies. All these challenges meant that farmers had to have financial resources, either their own or borrowed, to purchase equipment and supplies and to carry them through difficult periods. Settlers of small means often found the deck stacked against them.

In the course of developing ranches and farms, Californians learned to lean heavily on the state legislature. Bent on building a diversified economy, the California Legislature passed legislation establishing the California State Agricultural Society in 1854, its stated goals to encourage the development of agriculture and to foster the mechanic and household arts. To help promote these goals, the society was required to hold an annual public fair and livestock exhibition, at which awards for quality would be given. Some of the best-connected early agriculturalists in the state became members of the society's Board of Directors, among them the indefatigable Agoston Harasthy, who served as the president of the board for a year before making a European research trip to bring back grape cuttings for California. The initial issue of the *Transactions* of the State Agricultural Society (1858) included information on weather data–many newcomers were still trying to figure out the idiosyncrasies of regional climates–and Haraszthy's report on grape-growing and wine-making at his Buena Vista estate. Visiting committees also submitted detailed reports on farms ranging from Bidwell's Chico ranch and Sutter's Hock Farm in the north to converted mission lands and Wolfskill's citrus plantings in the south. In succeeding years the *Transactions* would report on many experiments in crop production under California conditions, but it would take decades before patterns would be established and optimism would grow about the state's agricultural potential.

Not far to the west of Sacramento, in Yolo County, a prosperous farm economy began to take root in the 1850s despite the early challenges. This area in the lower Sacramento Valley had several advantages from the beginning of statehood: the soils were deep and fertile along the several waterways that cut through it; markets were relatively close, either in Sacramento or in the San Francisco Bay area; and land titles were relatively easy to acquire, or so the first settlers thought.

In 1851 Champion I. Hutchinson and Charles E. Greene from Kenosha, Wisconsin bought their "Big Ranch" in Yolo County from the heir to the purported "Laguna de Santos Calle" Mexican land grant. This attractive property lay on the north side of Putah Creek, which ran into the valley from the mountains to the west. The ranch consisted of some 5,000 acres conveniently situated about twenty miles west of Sacramento. On the rich alluvial soil Hutchinson and Greene began to raise livestock and grow wheat, barley, and hay, going deeply into debt to develop the land, buy farm machinery, and harvest crops that were extensive for the time. As prices for meat, flour, eggs, and other farm products soared, the partners had great expectations for profits. Volatile weather, however, brought some serious flooding along Putah Creek, and other problems helped push the Big Ranch toward insolvency. The partners rented out some of the land for cash and sold portions of the ranch in 1856; then, mired in unpayable debt, they were forced to declare bankruptcy in 1858—a not uncommon experience for early landowners learning the hard way how to farm in California.

Along the length of Putah Creek and northward toward Willow Slough, a large number of squatters were also on the land by the mid 1850s. Land ownership in the area changed a number of times, especially after the Land Commission struck down the original Laguna de Santos Calle land grant in 1856, a decision upheld by the U.S. District Court in 1860. Like Hutchinson and Greene, some of the newcomers were from Wisconsin while a number of others came from Missouri. Most of them were experienced farmers, but they were still discovering how to adapt midwestern practices to California conditions, and they tended to concentrate mainly on livestock. The area became recognized for the quality of its animals, especially horses (which were in great demand) but several farmers also began to expand their cultivation of wheat as that crop became increasingly profitable. A few growers in favorable spots tried planting small market gardens or orchards that might bring a generous return from miners or city folk starved for fresh vegetables and fruit. Meanwhile, though Hutchinson and Greene continued to try to dig their way out from under their mountain of debt, the Big Ranch collapsed under an avalanche of lawsuits and was all but liquidated by 1861.[3]

One of Yolo County's early livestock men was young Hugh Glenn. In the summer of 1850 Glenn had returned to Paris, Missouri, with the money he had made from selling his Sacramento livery stable, but the bank in which he invested failed shortly afterwards. Having lost his capital, Glenn now had to consider his options. Missouri was prospering, and Glenn had good connections. His father ran a well-known hotel, but the son seems to have taken little interest in the business. He must have considered taking up the medical career that he had trained for, but he did not do this. Despite his education he seems never to have actively practiced medicine (although later in life he liked to be called "Doctor Glenn"). Instead, about the time Nancy presented him with their first child, Alphonso, in April 1851, he borrowed $6,000 to buy a number of cattle, horses, and mules to drive west, knowing that there were substantial profits to be made in California by shrewd dealers in livestock. After months on the trail and disposition of his animals in the Sacramento market, where he already had contacts, he took a job for the winter managing a small livestock ranch in the remote Stony Creek area midway up the Sacramento Valley. In 1852 he returned to Missouri, resolved to bring his family with him to California.

Thus in the spring of 1853 Glenn, his wife, and their young son journeyed west, again driving livestock for sale in California. The overland journey this time included Nancy's thoroughbred riding horse and a tin bathtub slung beneath their wagon for more civilized bathing than dips in the rivers and creeks they crossed. Back in California, Glenn and his family settled for a time in Yolo County along the north side of Putah Creek, where he raised livestock in partnership with his brother-in-law S.E. Wilson and his friend Marion Biggs, probably on rented acreage. All three young men hailed from Missouri's Monroe County, a region known for its quality animals, especially saddle horses, and the partners had strong links with the overland trade. In the spring of 1854 Glenn took his family up to the Stony Creek ranch to graze sheep while he also traded livestock. Thus, with two sojourns in the mid-Sacramento Valley in the 1850s, he became well acquainted with the possibilities of that region.[4]

Above Knights Landing, a steamboat stop on the Sacramento River, the valley was still a wide open country, framed by foothills and low mountains to both east and west, which grew in height as they curved together in the north. Americans had begun to explore the valley and miners were already trying their luck in mountain foothills, but little was definitely known about the area beyond Yolo County, and few had settled there. An early map of Northern California from San Francisco Bay to the Oregon border, prepared by U.S. Army topographical engineers in 1855, shows only blank space between Cache and Stony creeks west of the Sacramento River, while another map of the time marks "barren plains" there. John Bidwell, who had explored much of the valley in the 1840s, had seen

John Bidwell drew this map of the Sacramento Valley around 1850, showing the main waterways and land grants. The map shows large areas of marshland in the southern valley, relatively well watered areas to the north and east, and "sterile land" in mid-valley west of the Sacramento River. John Sutter lost his huge holdings over time, and the Rodriguez grant below the Rio de Capay (later called Stony Creek) eventually passed to Hugh Glenn.
Courtesy of The Bancroft Library, University of California, Berkeley.

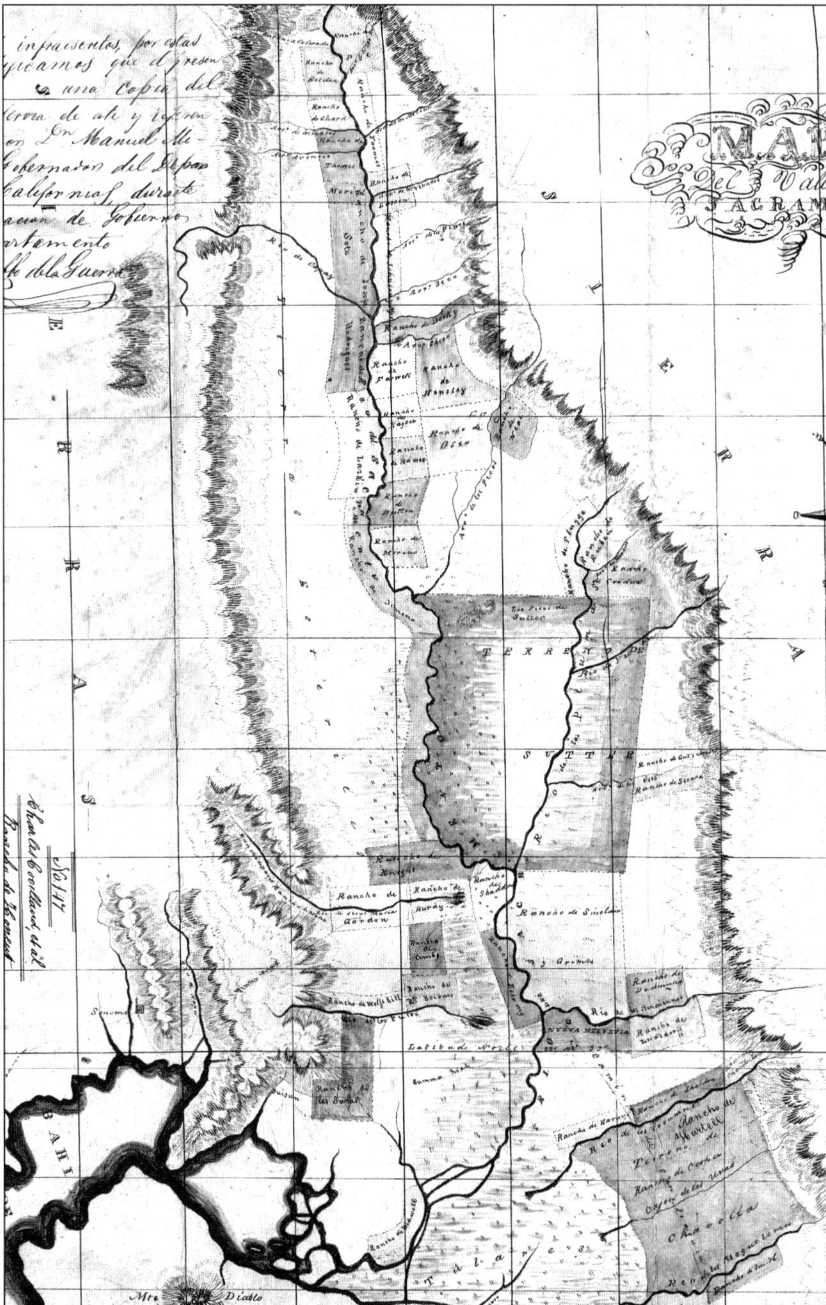

MAP
Del Valle
SACRAM
TERRENO DE
SUTTER
NUEVA HELVETIA
Rancho de Knight
Rancho de Gordon
Rancho de Sinclair
Rancho de Osio
Rancho de Hensley
Rancho de Moreno
Rancho de Los Berros
Rancho de Hartell
Chacolla
Latitud Norte
Sonoma
Mte Diablo

antelope by the thousands in the extensive grasslands and large numbers of grizzly bears in the riparian areas, but he himself finally chose to settle along Chico Creek to the north, where the natural resources seemed more promising for the development of the kind of diversified farming he wanted to pursue. In the middle of the valley, between the extensive marshes along the lower Sacramento River and the scattered woodlands bordering Stony Creek, there appeared to be little to attract permanent settlers. Most of the wide, dry mid-valley plains still lay untouched except by a few sheep men who grazed herds seasonally.

The first American settlers in the valley clung close to the Sacramento River itself, which offered greater potential than the land stretching away on either side. A broad band of mixed woodland followed the river, and massive trees festooned with wild grapevines provided shelter for a rich assortment of wildlife. The fertile overflow ground, created by centuries of periodic flooding, supported profusions of native grass. Yet while the riparian zone offered a striking contrast to the plains beyond, some of it was also a near-impenetrable jungle of thick vegetation and twisted vines, with downed trees, hillocks of debris left in the wake of winter floods, and quantities of gravel deposited in the low spots. Streamside settlers soon learned that the apparently placid Sacramento River could turn into a snarling enemy if heavy winter rains battered the mountains and foothills.

One main road ran up the valley, and that was the Sacramento-Mt. Shasta road near the river, which carried supplies to the mines and tiny communities in the northern foothills. Along this dirt road—muddy ruts in winter, a wide, dusty trail in summer—lay a few small grain ranches and rest houses at team-switching points, where the first settlers sometimes eked out a precarious living by serving occasional travelers. The small bands of Indians who still lived close to the river were gradually being pushed out, yet not many of the pioneer settlers or the prospectors searching for wealth in the mountains stayed long.

Here in the valley, near the curving thread of Stony Creek, the Glenns spent several seasons. The young family lived in a simple cabin built of willow poles and canvas, pioneer style. Isolated as it was, they probably enjoyed the adventure for a time, but Nancy's second baby, Emma, was born in 1854, and after months of primitive living and what must have been considerable loneliness for a well-bred Missouri woman, Hugh took his family back to Sacramento. The town was still filled with pack trains and miners collecting supplies for remote camps, but it was also a civilized community where social life, medical care, and transcontinental connections were available. By 1855, however, Sacramento was in something of an economic slump as placer mining declined, and in 1856 Glenn sold out his business interests to return once more to Missouri, resolved this time to stay there. Their third child, a son born that year, died in infancy, either on the return journey or after arriving back home. Young Emma too died in 1857 and was buried in Missouri.

Glenn's father was now the proprietor of the Glenn House in Paris, Missouri, a substantial three-story brick lodging built in 1855 on the site of the former log hotel. But the young Glenn was apparently not attracted toward a sedentary career in town. Torn between the needs of his family and his restless desire for the West, Hugh Glenn decided to become a cross-country livestock broker, making regular overland trips to bring improved breeds of livestock from Missouri and Kentucky into California. There was substantial money to be made in the western markets, for Californians were eager to improve the quality and supply of their animals, and practiced traders with good stock could name their prices. Though losses

Young Hugh Glenn was a handsome blue-eyed man with a taste for adventure as well as a talent for business.

Courtesy of Willows Museum.

could be heavy on the overland trails, the surviving animals could be sold for triple or quadruple their cost in Missouri, sometimes even more. These continental crossings became easier to make after 1858, when the federal government improved a portion of the Oregon Trail called the Lander Road, between South Pass and Fort Hall, in order to facilitate travel.

Glenn is said to have crossed the plains twenty-two times—11 round trips—to drive cattle, horses, and mules across the continent before the railroad made such drives obsolete. The journeys were always demanding, but Glenn was hardy and resourceful, with the added advantage of medical training. On one occasion Glenn was caught in a surprise Indian attack when he tried to retrieve some stolen horses, but managed to survive until he was rescued by his companions. Over the years he became well known in the livestock trade, bringing into California improved breeds of American cattle, Kentucky horses famous for spirit and speed, and the legendary Missouri mules. He made important contributions to California's rapidly improving livestock industry, and his talents caught the attention of influential backers.

Sometime around 1861, judging by the births of their children, Hugh and Nancy Glenn moved permanently back to California. Their third son, Charles, was born in Missouri in 1858, and their second daughter, Ella, in 1860; their fourth son, Frank, was born in 1862 in California, as were their succeeding children. Although Glenn continued his livestock drives across the continent for a few years longer, his family now settled into a home in Sacramento.

The year 1861 was a good time for the Glenns to stake their future on the West, for the beginning of the Civil War in April boded ill for border states like Missouri that were sharply divided in their views on slavery and secession. That year Missouri was one of the first to suffer through early skirmishes between Union and Confederate forces. Hugh Glenn, who himself reported owning a household slave in the 1860 Missouri census, may have judged that his family would be better off away from the conflict, but he was not the only Missourian to avoid the war. Many had already headed west in the 1850s, some of them settling in the Sacramento Valley, and as political and military tensions mounted, more set their sights there. One son of Missouri escaping the war was young Samuel Clemens from Hannibal, a river town not far from Glenn's own home in Monroe County. After a short stint in military service, Clemens chose not to re-enlist but jumped at the chance to accompany his brother to Nevada, where in the early 1860s he started his career as a journalist calling himself "Mark Twain" and spent several years working out of San Francisco, thus avoiding the grisly scenes of the War between the North and South.

An additional reason for Glenn to settle in California presented itself after 1860. The discovery of the Comstock Lode in the Nevada Territory revived the towns of Sacramento, Placerville, and Auburn as busy supply depots for packing personnel and equipment across the mountains into mining camps near Virginia City and the booming Carson Valley. The exploitation of the rich new mineral resources in the Comstock offered investors the prospect of extracting even larger fortunes than those accumulated by placer mining, and this time heavy-duty engineering changed the nature of western mining forever. Thousands of animals were needed for freighting innumerable tons of supplies, and brisk stage traffic sped into Washoe, as the territory was then known. An energetic livestock dealer could do very well in Sacramento in the early 1860s. Though the details of the Glenn family's residential moves are not known, in 1866 Glenn was listed in the Sacramento City Directory as a "stock trader," with his residence at 23 12th Street, just a few blocks north of the new state capitol building then under construction.

Rapidly growing San Francisco saw Henry Miller doggedly working his way up in the meat business in the 1850s. Within two years of opening his own butcher shop he felt confident enough to hire another man to run the shop from day to day while he rode horseback from the city south along the peninsula and east into the San Joaquin Valley to see what cattle were being raised on the scattered ranches there. He wanted to see where his meat was coming from, and he wanted to avoid the middlemen if he could.

His first journey, probably made in late winter or early spring, must have been an adventure. Passing southward through the beautiful Santa Clara Valley with its stands of great oak trees and luxurious expanses of grass, bounded on the west by the coastal range and redwood forests; he had to head east through Pacheco Pass in the Mt. Hamilton mountains. This was a tortuously narrow route but spectacular in its final view of the San Joaquin Valley–a panorama of 50 miles across the plains to the snow capped peaks of the Sierra Nevada. In the 1850s the isolated San Joaquin still had few travelers, the vistas were sweeping, and large bands of elk or wild horses could be seen grazing or moving through the grasslands. Miller scouted much of the northern part of the valley and soon grew to know it well, meeting the few early ranchers trying to eke out livings along the waterways. Always alert to possibilities, he could see that it might well be prime cattle country, and he filed his observations in his prodigious memory.

By this time some five dozen butchers were listed in the San Francisco city directory. Most had small shops, and competition was keen. In 1857 Charles Lux approached his acquaintance and competitor Henry Miller to propose a joint purchase of a big herd of cattle from Texas then being held at Pacheco Pass. Miller agreed, and they paid $67.50 a head for 1600 animals, American cattle that were far superior in quality to the stringy Mexican stock still being sold. This purchase price seemed enormous at the time, but the two young butchers were able to sell the animals in turn to others for a substantial profit. After this lucrative transaction Lux seriously considered cashing out his business and returning permanently to the east coast, but he soon changed his mind. The opportunities for two shrewd young Germans in California at the time were too promising to resist.

The original four-league Buri Buri grant owned by Isidor Sanchez and his wife had covered much of the southern peninsula from the San Bruno Mountains to what is now San Mateo. Charles Lux's Baden portion of the property, purchased in 1853, was very well located for business in nearby San Francisco. By the El Camino Real it was a seven-hour stagecoach trip from downtown San Francisco to the settlement at San Jose, with roadhouse stops along the way at regular intervals. Baden lay near the Twelve-Mile House along the tidal slough called Colma Creek, and the surrounding area was mostly farmland enjoying a salubrious climate along the sheltered bay. In 1857 Lux married Miranda Potter, a widow with a small son, whose first husband had been killed in a steamboat explosion. Lux and his wife soon built a comfortable home on the Baden estate, but meanwhile maintained a city residence in San Francisco.

In 1858 Lux and Miller, both now in their early thirties, decided to outpace their competitors by joining in a formal partnership that would include their own production of livestock as well as butchering and retail sales. Miller would take charge of the country or ranching end of the business, while the genial Lux would manage the San Francisco end, arranging for financing, purchasing, and sales. This turned out to be an ideal arrangement, utilizing each man's best

This early portrait of Henry Miller appears to have been the first of many created over his lifetime. Dressed in a fine wool suit, he seems eager to make a good impression. *Courtesy of Gilroy Museum.*

talents. Miller, who had already scouted the San Joaquin Valley, was the man in the saddle, the cattle broker and land buyer, while Lux was the partners' public face in the city. Thereafter known as Miller & Lux, the partnership throve from its beginnings.[5] The two men were not at all alike personally—Lux was tall and good-looking, with social skills that Miller lacked—but they got on amicably enough, and their partnership would last a quarter of a century.

The two men were in the right place at the right time. California's great cattle boom of the 1850s was a direct outgrowth of the Gold Rush. Prices had shot up from a mere $4–6 a head in 1846 to as much as $300 in Sacramento at the close of 1849, and even two years later cattle were bringing from $50 to $150, undreamt-of wealth to those who still had land and stock. Cattle had been the staple of the California economy for years, from the pastoral life of the Mexican ranchos to their conversion under Anglo impact. But the surging market for beef in the mining camps and growing towns now encouraged an unprecedented expansion of production, and for the next decade cattle sales constituted the chief source of revenue in the state after mining. The ragged Mexican herds were gradually improved by interbreeding with new American stock driven overland in great numbers from the Great Plains and Mississippi Valley by men like Hugh Glenn during the 1850s. Hundreds of thousands of animals were trailed over the roads into California, a reported 40,000 each year during the middle of the decade. By 1860 more than one million head grazed in California, more than double the number during the Mexican era.[6]

As the speculative mining economy began to stabilize with long-term investments in underground extraction, San Francisco entered a boom period. Urgent demand for meat in the thriving city and in equally prosperous Sacramento and the proliferating mines encouraged big cattle drives from the great Southern California ranches northward. These drives of a thousand or more head, herded by a trail boss and several vaqueros, might take many days, and predictably the cattle would walk off a certain percentage of their weight. Henry Miller began to believe that raising cattle closer to the northern markets made more sense. And he began to look for ways to acquire land for that purpose.

As Miller and Lux were positioning themselves for major expansion in their cattle business, almost immediately their partnership arrangement included family ties as well. In 1858 Henry Miller married Miranda Lux's sister, Nancy Wilmarth Sheldon. She lived only a year after the marriage, however, dying in childbirth along with the child. The following year, in business-like fashion—Miller is said to have proposed to his first wife's niece after a short acquaintance and given her two hours to make up her mind—the 32-year-old German married his second wife, 20-year-old Sarah Wilmarth Sheldon, by whom he would have four children over the following decade. Now Miller & Lux were established not only in business, but in society.

Miller began his decades-long strategic acquisition of property in 1859, when he purchased a portion of the Las Animas Rancho in southern Santa Clara County near the thriving town of Gilroy. Dozens of settlers had already discovered this fertile and beautiful country, and the Las Animas grant was being rapidly subdivided. Miller called his Gilroy property Bloomfield Farm and bought up more land nearby as it became available. (Over the years there would be extensive litigation over Las Animas property titles, involving hundreds of claimants in the Gilroy area, and Miller himself would pay for much of the legal work in order to help resolve ownership issues.) Bloomfield was well watered and had good grass, and it would become the central station in the partners' movement of cattle from the San Joaquin country to San Francisco. The 1,700-

acre property lay close to Pacheco Pass, the main corridor between the San Joaquin Valley and the coastal counties, and it provided an ideal resting point for cattle driven over the pass toward the San Francisco market.

Pacheco Pass, the gateway between the San Joaquin and Santa Clara valleys, transects a range of high hills west of the San Joaquin River, its route carved by streambeds flowing both east and west. Indian tribes used this pass for centuries to cross between the inland plains and the more verdant coastal valleys. In his explorations of interior California in 1805, Gabriel Moraga discovered a good camping spot in the pass, located where three Indian trails met near a pond of water. Moraga named this place San Luis Gonzaga for the saint's day on which he found it, and the spot became a well-known rest and recuperation area for travelers. In 1844 Juan Pacheco built an adobe hacienda here as the center of his 48,000-acre Mexican land grant. After his untimely death the grant went to his father, Francisco Perez Pacheco, who already owned the San Felipe grant to the west. The crossing in this area thus became known as Pacheco's Pass.

Later gold seekers traversed the trails of the pass on their way to the mines in the Sierra, some of them stopping at the Pacheco hacienda. Cattle and sheep were also herded through the pass as their owners moved them between valleys, but there was no real public roadway until Andrew Firebaugh, an American settler, began developing a toll road in 1856. In those days most local governments could not afford to build long-distance public roads and therefore sometimes granted private investors the right to create toll roads and charge travelers for their use, usually for a period of 10 to 20 years. Firebaugh's road project soon passed into other hands, but in 1858, when the way was opened, the Butterfield Overland Stage began using Pacheco Pass as a segment of its 2,800-mile southerly route from Missouri to San Francisco.[7]

In the early days of settlement the coastal side of Pacheco Pass and all the way up the San Mateo peninsula was—and still is—gorgeous country, especially in springtime. After favorable winter rains the scenery could be spectacular, with newly green hills providing a stunning backdrop for enormous waves of wildflowers. One ecstatic nineteenth-century woman wrote: "Here they have flowers in May, not shy, but rampant...flowers by the acre, flowers by the square mile, flowers as the visible carpet of an immense mountain wall. You can gather them in clumps, a dozen varieties at one pull. You can fill a bushel-basket in five minutes. And the colors are as charming as the numbers are profuse. Yellow, purple, violet, pink, and pied, are spread around you, now in separate level masses, now two or three combined in a swelling knoll, now intermixed in gorgeous confusion. Imagine yourself looking across to a hundred acres of wild meadow, stretching to the base of hills nearly two thousand feet high—the whole expanse swarming with little straw-colored wild sunflowers, orange poppies, squadrons of purple beauties, battalions of pink—and then the mountain, unbroken by tree or rock, glowing with the investiture of all these hues, softened and kneaded by distance." [8]

The flowers extended inland too. When John Muir traversed Pacheco Pass and the San Joaquin Valley on his way to the Sierra in April 1868, he too was struck by the immense sheets of color. "The Great Central Plain of California, during the months of March, April, and May, was one smooth, continuous bed of honey-bloom, so marvelously rich that, in walking from one end of it to the other, a distance of more than 400 miles, your foot would press about a hundred flowers at every step. Mints, gilias, nemophilas, castillejas, and innumerable compositae were so crowded together that, had ninety-nine percent of them been taken away, the plain would still have seemed to any but Californians extravagantly flowery.... one sheet of purple and gold, with the bright Sacramento pouring through the midst of it from the north, the San Joaquin from the south, and their many tributaries sweeping in at right angles from the mountains, dividing the plain into sections fringed with trees." [9]

Such scenes came during springtime in good years, of course, and even the engrossed Henry Miller must have enjoyed them in his treks across the pass over the years. But in summertime and during drought the great San Joaquin plain opening out beyond the pass was very different–dry, dusty, oppressively hot for weeks at a time. Between the scattered waterways, shrunken to thin lines of green, the vast stretches of native grass and wildflower residue would then lie parched and brown, dormant until the fickle rain began the cycle all over again.

Travelers responded in kind. William H. Brewer, who traveled the length and breadth of California between 1860 and 1864 in his work as a member of J.D. Whitney's State Geologic Survey team, climbed dozens of mountain peaks, traversed innumerable miles of varying terrain, and viewed spectacular vistas from the coast to the Nevada border. No one better documented the land of the state during these years than he, and his journal reads in part like a grand adventure. Yet his descriptions of the San Joaquin Valley–none in springtime–were not nearly as flattering as Muir's. "Oh!," wrote Brewer in his journal in October 1861. "What a tedious plain–league after league stretches away...desolate, without trees save along the river, without water nine or ten months of the year, and practically a desert." And the following June, despite the winter rains, he noted, "The soil and herbage is dry and brown, few green things cheer the eyes, no trees (save in the distance) vary that great expanse. Tens of thousands of cattle are feeding, but they are but specks save when they cluster in great herds near the water." When his party reached the eastern side of Pacheco's Pass, he noted the temperature of 102 and the high winds that poured over the pass into the valley. Two years later, in the middle of the killing drought that then gripped California, he described the San Joaquin Valley as a scene of dust, desolation, and dead cattle while his party passed southward toward Fresno, and then a landscape of monotony, mirages, and whirlwinds snaking across the plains leading toward the Kings River.[10]

It would not be an easy land to conquer, even for Henry Miller.

---

1 On the issue of Mexican land titles, see Hittell, *The Resources of California*, pp. 453–62; Robinson, *Land in California*, pp. 91–109; Warren, *California Ranchos and Farms.*

2 Eigenheer, Chapter 7.

3 Vaught, *After the Gold Rush*, describes early Yolo County farming in the Putah Creek area.

4 Sources on Glenn cited in Chapter 1.

5 See Igler, *Industrial Cowboys*, for a well researched though thesis-driven study of the Miller & Lux partnership.

6 Pulling, *History of California's Range Cattle Industry*, Chap. 1.

7 See *The Road Over* for collected reminiscences from users of the pass.

8 Quoted in Schmidt, "Through Other Eyes," *Fremontia*, April 2003.

9 Ibid.

10 Brewer, *Up and Down California*, pp. 202–3, 283, 379ff.

~3~

The Geometry of Settlement

THE CIVIL WAR THAT CONVULSED THE NATION from 1861 through 1865 had relatively little direct impact on faraway California, although California and Washoe gold helped support the Union by averting the wartime collapse of the government bond market. Some 16,000 Californians entered military service during the war, but most served within the state and missed the bloody battles in the South. War demand for food encouraged California farmers to plant wheat and thus helped fuel the growth of the western grain trade, but the general national economic problems related to huge diversions of money and supplies to the war effort were reinforced in the state of California, despite new mining ventures, by fiendishly difficult weather in the early 1860s.

Extraordinary amounts of rainfall and massive flooding along the rivers in the winter of 1861–62 turned the Central Valley into a vast inland sea.[1] Prolonged runoff from the mountains and foothills raised water levels in the sloughs that criss-crossed the sodden plains, spilling out to damage ranches and farms and drown an estimated 200,000 head of livestock. The floods cost hundreds of lives and millions of dollars, and created havoc in the new capital city of Sacramento, where 23 inches of rain fell in December and January and the two rivers bordering the city inundated the entire downtown. Residents had to move through the streets in boats, while flimsy structures were swept away and the lower floors of more substantial buildings were coated with mud. A series of these disastrous floods finally convinced Sacramento city authorities to haul in hundreds of thousands of cubic yards of dirt fill to raise the grade of twelve entire downtown blocks two feet above the high water mark. This effort took almost ten years, but the newly engineered riverbank, together with new levees upstream, offered significantly more protection to the business district during periods of high water.

Ironically, immediately after the 1862 flood came a two-year drought that became one of the worst such periods in California history. Although the high water of 1862 was followed by a luxuriant growth of vegetation that spring, the grass dried up as usual by the end of summer, and the winter rains did not return. The livestock that had grazed on the plains and foothills began to suffer. The first dry winter was threatening, but the second brought ruin. In the Sacramento Valley less than eight inches of rain fell throughout 1863–64, only a third of what might be expected in an average year, and it was worse further south.

Stock raising in California was at its height in 1861, when the state supported a couple of million head of cattle besides large numbers of horses, mules, and sheep. The three years of bad weather changed all that, especially in Southern California and the San Joaquin country. As ranchers tried to sell stock, cattle prices dropped to lows unheard of since the 1840s, and cattle were slaughtered for the negligible value of their hides and horns. Heavily indebted ranchers had no choice but to sell out. "The fall of 1863 was unusually dry; and even the winter months...brought no relief. Day after day went by with cloudless skies; and the grass failed to sprout from the famished earth. The springs and water holes dried up, and the great ranges were eaten bare of every kind of feed....The loss of cattle was fearful...the plains were strewn with their carcasses. In marshy places and around the Cienagas, where there was a vestige of green, the ground was covered with their skeletons, and the traveler for years afterward was often startled by coming suddenly on a veritable Golgotha–a place of skulls–the long horns standing out in defiant attitude, as if protecting the fleshless bones. It is said that 30,000 head of cattle died on the Stearns Ranchos alone. The great drought of 1863–64 put an end to cattle raising as the distinctive industry of Southern California."[2]

Weather that destroyed the southern cattle industry, however, turned out to be the making of Henry Miller and Charles Lux, who bought land that others were anxious to sell. The partners had already acquired significant holdings south of San Francisco with their combined 3,500 acres split between Lux's Baden and Miller's Gilroy properties, two sites strategically located for holding and feeding market-bound cattle. Now they moved decisively into the San Joaquin Valley.

In 1863 Miller & Lux bought 8,835 prime acres of the Rancho Sanjon de Santa Rita, level land lying along the west side of the San Joaquin River near a spot called Los Banos (the baths), a series of pools along a creek used by an early Franciscan missionary traveling through the valley. The partners bought the ranch from the well-known William Dunphy and Thomas Hildreth, pioneering cattlemen who had owned it only a few years after acquiring it from the heirs to a Mexican land grant. Although the land was well situated, the ranch buildings and corrals were still primitive, even forlorn, and the weather had made Hildreth anxious to sell out, thinking his fortune might be better in the mining districts. The $10,400 purchase price included all the livestock on the land plus the Hildreth & Murphy Double H ("HH") brand.

The Santa Rita ranch contained natural wetlands along the meandering course of the river and its tributaries, offering a usually reliable source of water except during extreme drought periods like this one. Miller & Lux suffered in the drought too, losing an appalling two thirds of the herd of 7,500 cattle they acquired with the land. Unlike other ranchers then being forced out of business, however, the partners had sufficient capital resources behind them to survive—a result of both Lux's amiable relations with city bankers and Miller's ability to make every nickel count. The two men took advantage of the depressed prices at the time and learned from the experience.

Miller & Lux bought more land as it became available, always at low prices, and held onto it as prices rose. Gradually the partners bought up all of the original Santa Rita land grant, averaging just $4 an acre for the entire 48,400 acres, and in 1868 they purchased a large part of the Orestimba grant downriver from the Santa Rita. The flat plains away from the river could be sun-baked and desolate during the long dry season, but there was green grass somewhere on the Santa Rita property nine months a year. After the drought broke and the rains returned, prices for market steers soared to $40 to $50 a head. Miller & Lux began to raise quantities of hay and kept stockpiles of it as insurance against the next fitful spells of weather. As their herds grew, they filled the pastures at the Santa Rita to capacity.

Miller & Lux had now acquired a set of properties that formed an extraordinary combination. The Santa Rita land lay a day's journey by horseback east of Pacheco Pass, and Miller's Bloomfield Farm about the same distance to the west. The Pacheco Pass toll road, which operated until the 1870s, had three rest areas in its 40-mile length, including the old hacienda and the San Luis campground. The road was supposed to be fifty feet wide, but it was mainly a one-way track, following the creek in some sections and very steep in others. From the Santa Rita ranch it generally took three nights to move cattle west across the pass to Bloomfield, then three more to get them north to San Francisco. Bloomfield had good pastures and extensive corrals, where the livestock could rest until being herded 70 miles onward to the Lux property at Baden. Baden was the final stop along the trail, where the tired animals could cool down from the journey before being driven in to city slaughterhouses.

In those early days Bloomfield lay conveniently close to the San Jose Stage Road that ran up the peninsula. This road served not only as the main thoroughfare for passenger and freight traffic, but as a major cattle trail up to the city in the 1860s. Before ranchers began shipping their cattle by rail, it was common to see herds of livestock being nudged along the road by half a dozen or more vaqueros. Soon, however, the railroad would follow the same route, with stops at both Bloomfield and Baden Station, and the Miller & Lux cattle would travel by rail car.

Buying land became Miller's passion. An excellent horseman, he spent days at a time investigating potential purchases, and he was a shrewd land buyer. Recognizing the vagaries of the California climate, he aimed at buying riparian properties and insisted on ironclad water rights, and his early choices were made at a time when there was relatively little competition for land in the Central Valley. Public land had not even become available until 1858, after state surveyors had sectionalized most of the public domain, and it was not until after the Civil War, in the late 1860s, that speculators began actively taking up sections in the San Joaquin Valley.[3] Miller & Lux were a little ahead of the game and knew exactly what they wanted.

Within a few years the partners owned a 40-mile stretch along the western side of the San Joaquin River and filed claims to more property south along the river nearly to Tulare Lake. To keep their carefully chosen land protected from other ranchers' livestock, they began in 1867 to fence their borders in an expensive years-long campaign. They also began expanding their holdings in the coastal counties by purchasing the Dequisquito Ranch in Monterey County (now San Benito County); some 24,000 acres of the Lomas Muerto Rancho in Santa Clara County, adjoining the Las Animas grant on the south; and the Peach Tree Ranch in Monterey County. These higher elevation ranches lay closer to the cool coast and offered more varied opportunities for grazing than land inland. In 1869, moving boldly toward the still largely open southern San Joaquin Valley, the partners began buying acreage in Kern and Tulare counties, again always with an eye on water rights.[4]

Miller & Lux bought out the heirs of grants; made loans to faltering farmers at the prevailing heavy rates of interest and foreclosed when the loans were not repaid; bought land scrip heavily from soldiers and Indian fighters who had received it in payment of service; bought swampland for $1.25 an acre in return for reclaiming the land with levees and ditches; and took full advantage of the loose land laws of the time by arranging for some of their employees to make "dummy" entries on government land, usually in 160-acre homestead parcels but sometimes more, which were then reconveyed to the partnership in local courthouses. Deed records in Merced County from the mid-1860s through 1887 show Miller & Lux as the designated grantee in 287 land transfers, most in 160-acre blocks.[5]

Miller's land-buying obsession was expensive, and his partner was not always enthusiastic, but the good-natured Lux maintained excellent relations with San Francisco financiers, and his friendships and social connections were very helpful in business. The headquarters Miller & Lux office was located on Kearny Street near other leading business firms. With personal access to bankers and investors, Lux seems seldom to have failed to come up with the money needed for Miller's purchases. He even learned how to manipulate the monetary markets when occasion presented, at one point importing gold from Germany to buy depreciated U.S. dollars to pay for property.

Like other large land buyers Miller & Lux attracted criticism as they accumulated enormous blocks of property. Unlike many other large buyers, however, they had little interest in speculation. They bought land for their own livestock production, and Miller was deeply involved in all the details of ranching. The partners held no mortgages and progressively made many improvements on their properties. Though they took advantage of circumstances, watching for new developments and pouncing when the time was right, they observed the land laws (lax as they were) and bought land from willing sellers. They did not cheat the government, nor did they cheat individuals. At times Miller exhibited genuine generosity toward settlers whom he respected. He believed that the firm would prosper in the long run only if it was fair and upright in dealing with others. Whether this was a moral or a strictly business principle, over the years it worked.

Profound change began to take place in the Central Valley after the Civil War, particularly in that portion north of the great Delta where the valley's two major rivers drain into San Francisco Bay. During the Mexican era wheat had been grown in California on a limited scale and in fact was even shipped in modest amounts during the 1830s to Russian settlements in Alaska, the Hudson's Bay Company in the Northwest, and the Sandwich Islands. But wheat production was greatly encouraged by Gold Rush demand for flour and bread. In 1850 California's entire wheat output was estimated at a scant 17,000 bushels—not even enough to feed the state's burgeoning population—but only two years later, encouraged by soaring market prices, the state crop was reported to be 297,000 bushels. By 1855 California was considered self-sufficient in wheat production. (Nevertheless, during the Gold Rush years barley output remained considerably higher than that of wheat because of the huge number of working animals that had to be fed.)

Rising international prices soon encouraged California wheat growing on a much larger scale. By 1860 the state's output rose to nearly six million bushels. By then nearly a hundred local flour mills were operating too, and flour was being exported as well. California was first mentioned as a wheat-producing state in the 1860 federal census, when it was reported to be growing three percent of the total U.S. crop. Within ten years that percentage doubled. California's wheat production began to surge during the Civil War, with exports at first going mostly to countries around the Pacific Rim and then to Europe. Land that had previously been used mainly for livestock range was increasingly converted into grain fields as farmers seizing on new opportunity learned from each year's experience.

California's remarkable nineteenth-century wheat era was contingent on a combination of economic factors that fell conveniently into place for a frontier state struggling to establish a permanent agriculture. Wheat was a perfect crop for the treeless plains of the Central Valley, which had formerly been viewed as unfavorable for farming. It did not require intensive cultivation, could be grown without irrigation, could be shipped long distances, and was not dependent on local markets but was rapidly becoming a staple of international trade. Strong population growth in British industrial cities and social upheaval affecting rural areas in Europe during the mid-century had led to demand for large quantities of grain imported from overseas. The disastrous Crimean war of 1854–56 had also created a need for major imports, and the Chinese demand for wheat flour was soaring. The Chicago Board of Trade, organized in 1848 by Chicago's grain merchants, helped to regulate the expanding sales of wheat from the American Midwest, but the California trade moved separately, and in the globalization of the grain trade there were some distinct advantages for both the California product and its timing.[6]

"California Club" was the original variety grown before the American occupation; it was a sturdy plant with a short head and relatively light yield. The first variety introduced by Americans was the Chili wheat, both white and red. White and red Australian wheat appeared around 1854. The red varieties soon disappeared, but the white ones persisted and became very popular with growers, commanding good prices in world markets. "California White Velvet" eventually became the preferred variety–fine in color and texture, sought after because it could keep for long months during storage, and good for milling especially choice flour. Bakers liked its excellent quality. When American grain began to be shipped to Europe, the dryness of the California crop also became a distinct advantage. California conditions produced wheat very suitable for shipping, with high gluten content. Its hardness protected it from weevil infestations, and it could withstand long sea voyages without deterioration.

The timing of California shipments, from late summer through fall, was advantageous as well, for ships entering the state in the spring and summer could return to their home ports with full cargoes instead of in-ballast. Planting took place from November through February, with no need for irrigation during the usually moist, mild winters of the inland valleys, plenty of sunshine in spring, and no rain to interrupt the harvest starting in June. As farm machinery evolved, it became possible to plant and mechanically harvest very large fields on big ranches, and the declining need for human labor cut down on costs. Thus the whiteness and plumpness of the wheat berry, the flintiness or dryness of the harvested product, and the timing and cheapness of production together allowed California wheat to compete very favorably in markets the world over.

After California's punishing drought broke in 1865 with rains that brought three straight bumper crops, the good weather encouraged farmers to plant tens of thousands of new acres. By 1868 a writer for Bret Harte's new *Overland Monthly* in San Francisco could enthuse: "As the harvest comes on, and valley after valley hastens to pour her tribute into the commerce of the State, the rivers are dotted with sails, and the winding channels among the tules are clouded with the smoke of numberless steamers threading their way through the sea of verdure, loaded with the rich produce of the land. From every part of the world fleets come to bear away the treasure, and we have become one of the world's great feeders."[7]

Grain mechants organized the San Francisco Produce Exchange in 1867. By then Isaac Friedlander had already become the key player in the California wheat trade. His towering figure was a familiar sight in the business district, where he operated on Sansome Street both as grain buyer and shipping broker through the port of San Francisco. The opening of the transatlantic cable and transcontinental telegraph in the early 1860s helped him enormously in keeping up with the worldwide trade, while at home Friedlander began to keep close tabs on the progress of each year's wheat crop through an intelligence system remarkable for its day. Using agents in various districts to gather information on weather and plantings, he would estimate the total prospective state crop and make commitments to purchase and ship loads of grain earlier than anyone else. His social connections with San Francisco financiers gave him ready access to capital, and, like other wealthy San Franciscans, he moved into the fashionable South Park district on Rincon Hill, where family mansions were rising in a display of new-found wealth based on California resources. He also began investing heavily in San Joaquin Valley land. Always the speculator, he could envision a bright future there for productive farmland.

Friedlander was not the only one making investments in agriculture. Claus Spreckels, the former grocer and beer brewer who was now in a financial position to choose his next enterprise carefully, turned his attention to the business of sugar, a staple in much demand. In 1864 he organized the Bay Sugar Refinery in a large building built along the water to receive and refine raw cane sugar imported from Hawaii. An obsessive worker and a master of details, he spent nearly a year in Germany learning the process of sugar refining from the ground up and eventually invented a new refining process that cut the time from days to hours. His first refinery becoming too small to meet the demand, Spreckels opened a larger facility at Potrero Point, calling it the California Sugar Refinery and later reorganizing it as the Western Sugar Refinery. Spreckels would become the prime mover in both the Hawaiian cane sugar and the California beet sugar business. Sophisticated, determined, and political, he would also become a leader in the overall economic development of California and eventually one of the richest men in America.[8]

In contrast, Agoston Haraszthy's career reached its peak in 1862, when his book *Grape Culture, Wines and Wine Making, with Notes upon Agriculture and Horticulture* was published by Harper in New York. This publication, still a classic, reported on his 1861 tour of Europe to investigate regional grape growing and winemaking there. In this whirlwind tour, sponsored by the State Legislature with the intent of bolstering the California wine industry, Haraszthy and his son Arpad visited France, Germany, Switzerland, Spain, Portugal, Italy, Greece, and Egypt, bringing back thousands of varietal grapevine cuttings as well as planting stock for olives, almonds, and other tree crops suitable for California. The trip was productive but expensive, and Haraszthy was never fully reimbursed for it.

Despite his talents and many contributions to California, Haraszthy found himself in dire financial straits in the mid-1860s, largely because of the appearance of the killing pest *phylloxera* in his vineyards in Sonoma County, a pest that he himself may have inadvertently introduced into California along with his European cuttings. In December 1866 he was also seriously injured in a distillery accident. He filed for a declaration of bankruptcy in September 1867 and decided to move to Nicaragua, then being touted as a new frontier, to make a fresh start. There he wrote travel articles for the *Alta California* and with his son and another partner took over a large sugar plantation near the port of Corinto, planning for the production of export rum. Then, in July 1869, at the age of 57 and under mysterious circumstances, he apparently drowned accidentally in a river and was either swept out to sea or eaten by an alligator.[9] His death, a distinct loss for California, was noted in the *New York Times*.

Despite the awful weather of the early 1860s, settlers continued trying to domesticate the lower Sacramento Valley. Yolo County facilities expanded rapidly as contracts were let for the construction of a county courthouse and jail as well as a public hospital in Woodland. A new bridge greatly improved transportation along the Tule Jake road that crossed the marshes to Sacramento, allowing year-round traffic for the first time. Yolo County's good soils and availability of water attracted new farmers to join the area's livestock growers, and some farm families became firmly established. As elsewhere, however, land speculation became a problem. The *Western Shore Gazetteer* directory issued for Yolo County in 1870 pessimistically declared that land-buying opportunities were slim: "The curse of land monopoly has destroyed the hopes of the immigrants, who, if well informed, know that they cannot hope to obtain land here on terms that will compare with the inducements offered them elsewhere. ...the majority of the available lands are held by speculators, at prices far beyond the reach of a poor man." [10]

Increasing land prices in Yolo County, not to mention the general muddle of land title litigation along Putah Creek, must have influenced Hugh Glenn in his choices. Glenn ended his cross-country trips in 1865 to enter into a new partnership in livestock and grain with his old friend Marion Biggs, again along Putah Creek, apparently on rented land. The partners' venture was based about three miles west of Davisville, a new town being planned along the railroad tracks that were slated to run toward San Francisco Bay from Sacramento. There is no record that either partner actually owned land in Yolo County, and the partnership lasted only two or three years, for in 1867 Glenn began his own ranching enterprise in Colusa County to the north, and Biggs was elected to the state Assembly from Sacramento County.

Land prices may have seemed relatively high at the time in Yolo County, which lay close to the capital city of Sacramento, but in the more remote areas of the Sacramento Valley, as in the San Joaquin, land was still cheap after the great drought, and there were not many buyers because the region was still relatively unknown. On the plains midway up the Sacramento River, between Cache Creek and Stony Creek, it was no place for the easily discouraged. To the west of the river, dry, treeless vistas opened out toward the foothills, except for the one small natural pond between the river and the hills where the town of Willows would later spring up. In the foothills clover and wild oats might offer fine spring forage for cattle and oxen as well as elk and antelope, but in summer the unrelenting sun beat down and dried out everything. Away from the river, the soils of the plains were often heavy and difficult to handle, and desiccating winds could leave much of the land brown and crisp. To the north were only fringes of green along Stony Creek, which could roar like a lion in winter floods and then shrink to pathetic puddles between gravel bars in drought.

On the east side of the Sacramento River, in contrast, lay the Sutter Buttes to the south—those oddly isolated volcanic outcroppings that form landmarks visible for miles in clear weather—surrounded by miles of marshland. Here the marshes were home to swarms of mosquitoes in summer and to millions of migratory waterfowl in winter, huge flocks feeding along the water until spring's relentless pull northward.

Glenn had already spent several seasons working and living in the Stony Creek area, and he knew the region west of the river above the town of Colusa. In 1864, beginning his transition from livestock broker to landowner, he purchased four sections of land along the river near the hamlet of Princeton. In December 1867 he was able to make a decisive move in acquiring more property, a substantial 7,000 acres on the north end of the former Jacinto land grant, for which he paid $11,200 or just $1.60 an acre. According to a Glenn family story, Hugh, an avid poker player, won $10,000 at a card game in Sacramento, which became his main stake for that big land buy. The day after his win, he is said to have headed upstream on a river steamer to the Jacinto landing, where he nailed a horseshoe to a large oak overlooking the stream. The family regarded that horseshoe as his good luck charm; it stayed in place for years until the tree was undercut by flooding and fell into the water.

Glenn must also have had access to capital from Sacramento lenders, because almost immediately he bought up the 8,000 acres lying west of today's city of Orland, again at $1.60 an acre, for a total of $12,800. All of this land was part of the eight-league Jacinto Rancho granted in 1844 to Jacinto Rodriguez, which had passed three years later to the American W.H. McKee, who had it officially surveyed in 1850. Although its boundaries were litigated, McKee's claim for 35,488 acres was patented in September 1859 by the U.S. District Court. McKee gradually sold portions of the property starting at 50 cents an acre. Glenn's acquisitions encompassed the heart of the grant along the river, ensuring ready access to transportation as well as to water supplies, and containing valuable riparian resources.

Glenn acquired the owner's livestock along with the land, and his sale of the stock lowered the real cost of the purchase. When he began cutting some of the riparian woodland to clear the land for farming, sale of cords of firewood to steamboat companies and settlers helped to defray the expense of development. Some of the giant oaks near the river at the time were six feet in diameter and up to 70 feet tall, and the firewood brought good prices. (Some ranches along the river actually became known as "wood ranches" for their principal source of income, and landowners paid little attention at the time to the environmental implications of clearing the riparian zone.)

Now in his early 40s, Glenn was ready to go into farming in a big way, and he began using his initial land purchases as collateral for other large loans from Sacramento banks. The following year he bought another 6,000 acres to the west for only $1 an acre. In September 1868 Glenn and Biggs liquidated their partnership in Yolo County with a large sale to close out their operations there. Biggs himself soon also moved north, to Butte County above the Sutter Buttes, where he began growing wheat around Gridley. He maintained his friendship with Glenn and served as an influential contact on the State Board of Agriculture and later as a delegate to the state constitutional convention of 1878. Years later Biggs would also serve two terms in the U.S. House of Representatives.

Glenn had started out as a livestock man, and at first he must have managed much of his new ranch as a stock farm, competing with the wildlife on the plains or in the woodlands along the river. But year by year and field by field he cleared the land and planted grain, encouraged by the fertility of the virgin soil and the rapidly growing market for wheat. Land close to the river produced excellent wheat crops, from 40 to 60 bushels an acre, but Glenn soon found that even the heavier soils of the plains west of the river could, with favorable weather and good management, yield a respectable 25 bushels. "Dr. Glenn's first purchase [included] red, gravelly land, and when he began to plow it, nearly every one predicted a failure," wrote Will S. Green in his Colusa County history of 1880, "but it proved to be about the best wheat land in the county."

After 1868 the growing Glenn family lived for a time in the original rambling ranch house at Jacinto, building additions as needed and bringing in furnishings by riverboat. For a while they retained a home in Sacramento too. Their third daughter, Eva, born in 1864, was joined by two more sons in 1868 and 1870. While Nancy cared for the children with the help of a Chinese cook and a household servant, Glenn poured his energies into improving the ranch and buying adjacent land at increasing prices. With money behind him and the prospect of handsome profits ahead, he developed fields, erected buildings, and bought machinery. He would ride the crest of one of the biggest booms that California agriculture has known.

Riverboats provided basic transportation when Glenn began his career at Jacinto. The roads in the upper part of the Sacramento Valley were still primitive, stage routes were few, and railroads were in their infancy. The two stage lines running out of Sacramento on the east and west sides of the river as far as Shasta took about 30 hours or as many as three days to make the 140-mile trip, and the rides were grueling. The six-horse teams had to be switched every twelve miles, and with up to 28 passengers riding inside and outside the coaches, conditions were cramped and uncomfortable, especially when the way was dusty. Passengers often had to get out of the coaches because of obstacles along the road and found that for their fare they ended up walking part of the way.

The Sacramento River, on the other hand, was a natural highway from the great Delta to its upper reaches, and steamboats offered a far more pleasant journey than stagecoaches, even if they too had challenges. One challenge was the variability of river conditions. Before twentieth-century reclamation and flood control projects, the river in winter and spring carried immense amounts of runoff from the mountains to the Delta, while in late summer the stream turned placid and warm, shallow enough to ford in some places. River flow controlled traffic, but for most of the year in the first decades of statehood, steamboats supported the economy and the lives of those living within reach of the water. Farmers tried to locate near navigable waters for shipping their products to market, and steamboat owners competed for trade along the river. After 1854 the California Steam Navigation Company dominated most of the river business, but various small independent trading boats also operated up and down the river, making dozens of stops at tiny docks or landings constructed largely from brush.

The early California steamboats were a blend of small eastern and Mississippi paddlewheel types adapted to local conditions, with shallow drafts. Steamboats going upstream carried mining supplies and equipment, farm machinery and tools, building materials, and household goods. Downstream the vessels carried the products that

The nineteenth-century Sacramento River offered lovely water views as well as a convenient channel for steamboats plying between the landings scattered from the Delta to the head of navigation at Red Bluff. This lithograph appeared in the April 21, 1877 issue of the WASP, a weekly journal published in San Francisco.
Courtesy of California Historical Society, North Baker Research Library, San Francisco.

paid for all those goods: gold, livestock, poultry, and produce. Sometimes they pulled flat barges loaded with bulky nonperishable goods like lumber and grain. Larger steamboats also carried an eclectic mix of passengers, including miners, merchants, farmers, families, and the "bindle stiffs" or farm laborers who appeared seasonally for planting and harvesting grain. Passengers could sleep overnight in staterooms or in the main lounge and eat in the ship's dining room, watching by day as the boats pushed north through the walls of vegetation that still lined much of the river. In summertime many of the trees were draped with wild grapevines to the water's edge, lending an almost tropical appearance, but occasionally passengers could catch glimpses of clearings with farm buildings as the boats stopped to exchange cargo at private landings.[11]

One steamboat landing now long vanished but well known in its day was at Maine Prairie on Cache Slough, a tidal channel extending 16 miles northwest from the Sacramento River a few miles above Rio Vista and about 20 miles south of Putah Creek. Starting about 1860, the Maine Prairie landing was a bustling place for well over a decade, receiving loads of lumber and supplies while shipping out tens of thousands of sacks of wheat annually, perhaps more than from any other landing in the valley. Hauled in daily from the surrounding grain ranches by eight-mule-team wagons during the late summer and fall season, the wheat went out to San Francisco Bay flour mills or, increasingly, into the export trade. Partially financed by grain broker Isaac Friedlander, several warehouses and hotels did a thriving business along the Maine Prairie landing until the traffic there was eclipsed in the 1870s by the more reliable, though more expensive, railroad connecting Sacramento to Vallejo.

River traffic always had its perils. The winding Sacramento River offered significant challenges for navigators in its course down the valley, including shifting sand and gravel bars and thickets of flood debris piled up in numerous bends. Depending on the time of year, tule fog, treacherous currents, snags, or low water were common hazards, and throughout the year steamboat accidents were not uncommon as boats hit obstacles or boilers blew up. Some explosions killed both crew members and passengers. Sometimes boats would collide. Some boats sank after being damaged on snags, even on their first trips up the river.

Seasonal conditions always influenced travel. During the winter the upper valley could be isolated by rushing waters that stymied navigation. Later in the year, mucky shallows might block access to portions of the shore. In summer low water could pose the threat of grounding. Until 1852 the head of navigation reached only as far as the town of Colusa, but then a man named U.P. Monroe spent $50,000, a small fortune, to clear the channel up to Red Bluff—and subsequently went broke. Until 1871 boats could still reach Red Bluff, though only during seasons of sufficient flow and favorable weather. But navigation along the Sacramento began to deteriorate permanently as hydraulic mining in the foothills steadily emptied silt and gravel into the watershed.

Though the steamboat era lasted nearly half a century, it was in its heyday from the 1850s through the 1870s. As long as it lasted, river transport was relatively inexpensive and often colorful, offering economical shipping and one of that era's pleasures for travelers.

The geometry of settlement in California was about to change dramatically. Where the natural curves and bends of waterways had early dictated the movement of traffic and the location of communities across America, new settlement after the Civil War was influenced by the placement of railroad lines. In the great track-laying era that both preceded and followed the opening of the transcontinental railroad, railroad engineers learned to tame the landscape, building narrow trails for iron horses—fixed straight lines across plains and deserts and graded ascents through canyons and mountains. During the 1840s railroad tracks unfolded steadily across America's northeastern states, then across the Midwest. In 1853 a railroad line was completed even through the challenging terrain of the forty-mile Isthmus of Panama, providing far more convenient transit from the Atlantic to the Pacific coast. The daunting distance and huge expense of rail construction, however, as well as the sectional differences in Congress, kept the western half of North America behind.

Even so, short-line railroads began to link up towns and provide access to new places in California even before the Civil War. In 1854 the president of the proposed Sacramento Valley Line enlisted a brilliant young railroad engineer named Theodore Judah to take on the job of planning and building the first railroad west of the Mississippi River. Judah, born in Connecticut in 1826 as the son of an Episcopal minister, studied civil engineering at the renowned Rensselaer Technical Institute in New York and embarked early on a practical career that brought him recognition, before age 30, as a doer and builder in the rapidly developing field of railroad design. He jumped at the chance to be the pioneering railroad engineer of the Pacific Coast. By 1856 the short line between Sacramento and Folsom was completed, but Judah meanwhile developed a far grander vision of building a great railroad to run from Sacramento through the imposing Sierra Nevada to link with another line coming through the continent.

The talk of a transcontinental railroad had been in the air for years, and federal surveys had already identified five possible routes, but little more had been accomplished. Judah now wrote a long paper about his proposal that sounded a stirring call for action. His prospectus, "A Practical Plan for Building the Pacific Railroad," dated January 1, 1857, was meant "to inspire the capitalist with confidence," for Judah was skeptical about the ability of government to undertake such a large venture with any efficiency and believed that private investors should manage the effort. He began to promote the project with anyone who would listen. His enthusiasm earned him the nickname "Crazy Judah," but the California legislature agreed to convene a Pacific Railroad convention in 1859 and delegated Judah to lobby in Washington, D.C. for federal support. Gridlock in the capital, then choked by the North-South discord, thwarted any action at the time.

In 1860 Judah was invited by Dutch Flat storekeeper Daniel W. Strong to consider a route for the proposed line through Emigrant Gap and the Donner Pass, breaching what had been perceived as the almost insurmountable barrier of the Sierra Nevada. Excited by his discoveries, Judah drew up articles of association for a Central Pacific Railroad and began contacting potential investors. To his chagrin, San Francisco capitalists were not interested. He then appealed to four prosperous Sacramento merchants—Charles Crocker, Mark Hopkins, Collis Huntington, and Leland Stanford—in a meeting at Huntington and Hopkins' store at 54 K Street. The four agreed to invest $1,500 each for an official surveying party to be led by Judah in April 1861. A month later the young engineer presented them with a scaled Sierra route map some 90 feet long. In June, only weeks after the outbreak of the War Between the States, the Central Pacific Railroad of California was formally incorporated. The four merchants were designated as chief officers and Judah as the chief engineer. In October Judah and Huntington went to Washington D.C. with the map and the promise of financial backing to try to persuade Congress to lend its support.

The result of their lobbying was the Pacific Railroad Act of 1862, signed by President Abraham Lincoln in July. The legislation approved the building of both eastward and westward lines, with significant federal subsidies of land grants along the tracks for the companies that would undertake the mammoth efforts. By this time one of Judah's four backers, Leland Stanford, had been elected governor of California, and he used all his influence to advance the agenda of the new railroad.

In January 1863 the Central Pacific Railroad began construction activities in Sacramento with great fanfare. As construction pushed east, however, Judah became alienated by the partners' cost-cutting measures while they grew irritated with his apparent engineering errors. Railroad building demanded more and more capital, and the company faced severe financial stress. The four officers began to try to squeeze Judah, who had invested little money of his own, out of the partnership. In October 1863 Judah, feeling that his personal honor was at stake and bitterly at odds with the Sacramento four, sailed for New York to try to find financial backers who might help him buy them out. In Panama, however, Judah was stricken with yellow fever. He died shortly after he reached New York, not yet 40 years old, unable to see the completion of his dream.

The "Big Four," as they became known, afterwards took the credit for the dream—and the huge profits eventually generated. Despite enormous difficulties, Judah's hope for a transcontinental line through California was fulfilled at Promontory Summit, Utah, after a tumultuous competition between companies building from both east and west. On May 10, 1869, a Golden Spike driven into the rails linked the Central Pacific with the Union Pacific. The journey across the continent that had once taken six months could now be made in six days.[12]

A new era in the state's history thus began after 1869. Twenty years after the Gold Rush, travel to distant California was revolutionized. The transcontinental railroad carried in crowds of new immigrants and yet another set of speculators eager to exploit California's resources. The great artery of the new railroad was soon thickened and strengthened by the widespread extension of associated lines unfolding into all sections of the country. In California railroad track mileage in the Central Valley expanded exponentially in the 1870s and 1880s. Railroad building brought an end to the isolation of many land-locked areas, allowing the further development of farming and encouraging the growth of new settlements. Commerce advanced along with the iron tracks. Railway stops became railway towns, and the opening of both government and railroad lands brought rapidly increasing population to the interior valleys as well as along the coast, encouraged by vigorous railroad publicity.[13]

The pioneering short line between Sacramento and Folsom encouraged other short line proposals between the Carquinez Straits and the lower valley as far as Marysville, though none were successful before the Civil War. In 1865, however, with the Central Pacific working its way east, the California Pacific Railroad company began building a line from Vallejo toward Sacramento and to the north as far as Yuba City. To gain control of competing river traffic this company acquired the California Steam Navigation Company in 1871, but was itself taken over by the Central Pacific the same year. Other branch lines began to grow across the valley like a spider's web. On the west side of the Sacramento Valley, the Vaca Valley and Clear Lake Railroad inched northward in the late 1870s, while a couple of short lines struggled to gain a footing on the east side across the flood-prone Sutter Basin, with investors engaging in real estate speculation along the routes.

The east-west transcontinental railway effort naturally encouraged interest in a north-south trans-Pacific Coast line. The California and Oregon Railroad, organized in 1865, pushed gradually up the middle of the Sacramento Valley after 1868, reaching as far as the newly organized town of Redding in 1872. Redding remained the railhead for the next ten years, however, because the Oregon portion of the line pushing south collapsed in financial disarray after its tracks reached Roseburg, leaving unbroken the mountain barrier between Oregon and California. Despite its temporarily unrealized connection to the north, the new railroad changed the economy of the valley in immediate and impressive ways. The expansion of railroad lines served grain farmers well, for wheat could now be grown at some distance away from the rivers, without the burden of long wagon trips to navigable water. Formerly remote land in sleepy locations became newly attractive to settlers, and sales and prices soared as thousands of new acres were brought under the plow.

Some of the communities that had served river traffic began to fade away, while others rose along the routes of the railroads, the stations serving as the nuclei for small towns. One such settlement was The Willows, a site near the half-mile-long natural pond surrounded by willow trees that had served for years as a landmark watering place for stockmen in the mid-valley between Cache Creek and Stony Creek. This place began to develop into a community in the early 1870s, but grew rapidly after the railroad station opened in 1878. After a post office was established in 1879, the town became known simply as Willows.

Wheat became the foundation of the Colusa County economy. Though barge traffic on the river continued to move hundreds of thousands of sacks seasonally to market, growers also began to ship by rail as soon as lines became operational. Glenn's former partner Marion Biggs was the first to ship wheat from the depot subsequently named after him in Butte County; and it was reported that by 1880 Willows shipped thirty thousand tons a year by rail, much of it produced on the great Glenn ranch.

Railroads and development in the San Joaquin Valley were not far behind. The construction of rail lines there opened up whole new possibilities for this heretofore largely ignored region. At first the interior valley had seemed too remote for the railroads that developed along the coast south of San Francisco. A Southern Pacific Railroad company was incorporated in December 1865 to route a line from San Francisco to San Diego, with a section planned to run through Tulare County into Los Angeles, but construction faltered. In the meantime a small railroad that began to run from San Francisco to San Jose was acquired by the Southern Pacific and extended to Gilroy. In 1868, just before completion of the transcontinental railroad, the Big Four's rapidly growing Central Pacific took over the Southern Pacific. With the transcontinental connection achieved, the investors turned their ambitions toward the San Joaquin Valley, laying lines rapidly southward from a railway town called Lathrop (named to honor Stanford's wife) past new town sites of Modesto, Merced, Fresno, and others, to the point at which the line crossed the originally proposed route. This spot was called Goshen, located just east of Visalia, reached in the summer of 1872. Train service began almost immediately as rail sections were completed, and by 1874 the railroad extended to Bakersfield.

Before the coming of the railroad the broad plains of the San Joaquin were only sparsely dotted with a few lonely livestock ranches. Leland Stanford is said to have ridden through the valley himself on horseback and personally selected most of the station sites, and these places "were born as railroad towns and owe their development to that new mode of transportation." [14] Routing the line through the San Joaquin instead of along the coast undoubtedly made engineering easier, but another reason for doing so was that the company could locate its land grants along the right-of-way on government land still open there. Such subsidies were very important to the investors, and the railroad would take great pains to attract settlers onto these lands. Over the years the valley and the railroad would develop together.

Throughout the late 1860s and 1870s the Central Pacific rapidly absorbed most of California's short lines as well as several steamship and ferry companies. In 1873 the Big Four moved their corporate offices from Sacramento to a handsome new building in downtown San Francisco, using their personal wealth to build elaborate mansions on Nob Hill. The Central Pacific was merged with its other railroad acquisitions into a holding company called the Southern Pacific Company in 1885.[15]

The economic power of the Southern Pacific in the late nineteenth and early twentieth century can hardly be overstated, for it exercised a virtual monopoly over nearly all long-distance transportation in California. Steamboat traffic on the rivers was at best seasonally limited, and by the 1890s was beginning to decline. Motorized vehicles did not become widely available until after Ford's Model-T automobiles went into mass production around 1908, and no state highway system existed until after 1909.

Development in the Central Valley was shaped by railroad construction, but so too was the growth of some of the cities around San Francisco Bay. The city of Oakland became the western terminus for the transcontinental railroad in 1869. Here cross-country land transportation linked up with bay shipping and ferry service to San Francisco.

In 1851 the site of Oakland (then known as Contra Costa or the "opposite shore") had been described by one observer as an "evil-smelling strip of muddy tidelands on one side and a flotsam-filled creek on the other," but this at first unprepossessing place had fallen under the control of Horace W. Carpentier, a sharp operator who more or less swindled Don Luis Maria Peralta out of the shoreline part of his original land grant and then got the tract incorporated in 1852 as the city of Oakland. Carpentier became the first mayor of Oakland, and he and his associates were vigorous developers of the Oakland waterfront. He is also credited with the negotiations that finally brought the railroad there, profiting greatly by the connection.[16]

Oakland flourished after the arrival of the railroad, reconfiguring its marsh and tidal flats to create a real waterfront and experiencing rapid growth in manufacturing and shipping. During the 1870s a building boom transformed the city. With its rail and sea connections Oakland became the hub for shipping agricultural and other products along the entire west coast. Factories multiplied to process commodities and manufacture goods. Lumber mills, flour mills, breweries and canneries, sugar refining plants, cotton and jute processing facilities, and even carriage and wagon building businesses located near the improved waterfront. Oakland was "modern" too, with gas lights downtown, communication by telegraph and telephone, and horse-drawn streetcars and steam trains.

Early Oakland leaders founded several highly respected educational institutions. The town began to call itself the "Athens of the West" for its large number of schools, some of which attracted boarding students from several states. Female academies were prominent in the mix. Holy Name nuns arrived from Canada in 1868 to found a Catholic convent and girls' school on property overlooking Lake Merritt; and Mills Seminary for Women moved to Oakland from Benicia in 1871, becoming chartered as a women's college in 1885. Even the University of California, established in 1868 as the state's land grant college under the Morrill Act, began in Oakland, grafted onto the former College of California. (The University soon moved north to Berkeley.)

By the mid-1870s Oakland was the second largest city on the Pacific coast after San Francisco, with luxury hotels, wealthy residential neighborhoods, and working-class suburbs. Many grand homes were built around Lake Merritt, a reservoir formed by damming the tidal slough. Easy access to San Francisco by ferry helped Oakland become that city's bedchamber and a desirable address for well-off families. The climate was warmer and more pleasant than San Francisco's, with less wind and fog, and some families thought Oakland more suitable for raising children, away from urban emphasis on money and high living.

Oakland became a desirable address for the dead as well. The hills sloping upward from the bay boasted splendid views across the water and were punctuated with live oaks and seasonal wildflowers. In 1863 Mountain View Cemetery was established in the hills as a spacious and beautiful resting place for the famous and not-so-famous dead of the entire San Francisco Bay area, its landscaping designed by Frederick Law Olmsted. Mountain View's memorial monuments soon competed in grandiosity, reflecting Victorian tastes. Many of California's most prominent figures of the nineteenth century would be interred there–including Hugh Glenn and Charles Lux.

By 1870 California was poised on the brink of tremendous change, with new modes of transportation and communication shattering much of the isolation that had previously hindered the state from full participation in national politics and economic life. The 1870s would include great periods of turmoil both for the nation and the state, but California would continue to flex its new muscles, aim at broader development, and struggle for order. Part of the foundation for hopefulness would lie in the phenomenal rise of the international wheat trade, propelling state farmers into a far broader world.

Central Valley farmers of the nineteenth century discovered that wheat was an ideal cash crop. There would be a huge increase in production each decade for the next 30 years. Like much else along this far-off coast, wheat growing took off in bonanza style when men thought they might strike it rich. There were risks, of course: drought, destructive winds, untimely rains, field fires, unreliable transportation, even the manipulated markets. In a basically one-crop economy, all local business depended on the weather and the progress of the wheat crop. But for a time the good years outweighed the poor ones, and wheat was the gold that sprang from the soil.

---

1 County histories describe the terrible weather of the 1860s, as do McGowan, *History of the Sacramento Valley,* and Kelley, *Battling the Inland Sea.*

2 Cleland, *The March of Industry*, p. 61.

3 Eigenheer, pp. 312–18.

4 This narrative of Miller's land buying derives from the Miller dictation to Edwin Fowler; Lawrence, *Henry Miller and the San Joaquin Valley;* Igler, *Industrial Cowboys*; and Treadwell, *The Cattle King.*

5 Lawrence, ibid.

6 California's 19th-century wheat trade is described in two articles by Horace Davis in the *Overland Monthly* (1868 and 1898) and *Breadbasket of the World*, edited by Hikert and Lewis.

7 Davis, "Wheat in California," *Overland Monthly*, 1868.

8 For a concise description of Claus Spreckels' career, see Kirk, *A Flier in Oil,* a book mainly about one of Spreckels' sons.

9 See McGinty, *Strong Wine.*

10 *The Western Shore Gazetteer and Commercial Directory for the State of California:* Yolo County, Woodland, 1870.

11 White, *Glenn County Sketchbook,* includes a section on riverboats, and river traffic is described in detail by McGowan, pp. 71–85, 301–11.

12 Ambrose, *Nothing Like It in the World*, recounts Judah's story and that of the Big Four.

13 The history of railroads in the Central Valley is covered in McGowan, pp. 207–20, and Smith, *Garden of the Sun*, pp. 325–45.

14 Smith, p. 331.

15 See Orsi, *Sunset Limited,* for the history of the Southern Pacific.

16 Conmy, *The Beginnings of Oakland*, covers early city history.

~4~
The Gamble of Grain

A GOLDEN TIDE OF WHEAT was flowing out of California by 1870. Grain growing took off rapidly after generous rains broke the mid-1860s drought, but other factors were important too. Titles to the former Mexican land grants were now mostly settled; farmers felt more secure in making investments; and the opening of new areas of virgin soil coincided with a series of poor harvests in Europe, triggering market demand that reached halfway around the world. Rising demand also coincided with the quickening decline of California's free-range era, for in 1872 the legislature passed the "No Fence" law overturning the former dominance of cattlemen in the open countryside by making them newly liable for crop damages due to roaming animals. This blow to stockmen was a boon for farmers, who no longer had to build fences to keep wandering livestock out of their fields. They responded by cultivating more land. Wheat was soon being produced on more than a million acres.

California exports rose quickly, and international wheat prices, influenced by several European wars and by strong demand also from China, remained relatively high—usually well above $1.50 a cental (100-pound sack)—for a period of at least ten years. Wheat returns during the 1870s were usually quite favorable for growers, and they were especially good in 1873–74 and 1877–78, when crops were short because of poor weather. So important was grain growing that even the little Colusa weekly newspaper printed Liverpool wheat prices, though they lagged a few days behind the trade. In 1879 the Colusa County assessor reported 2.3 million bushels of wheat harvested from 220,500 acres, and the price that October rose to $2.05 under the stimulus of the Turkish-Russian war. Even after prices began to drop in the 1880s, about two thirds of California's wheat crop continued to flow into foreign markets. These bonanza years for growers served as the foundation for enormous development of California's agricultural resources. For well over a generation wheat was the essential underpinning of the Central Valley economy.[1]

The wheat grower's active seasons extended from November into mid-summer. Seed was planted in late fall to sprout and grow during the rainy season from December through March. The crop began to ripen in April or May and was harvested in June and July, then shipped throughout late summer and fall to distant ports in England, Europe, and Asia. Farmers learned to improve yields by fallowing some of their fields in alternate years, a water-conserving practice that was recommended early by the State Agricultural Society but did not become widespread until the 1870s. The fields to be fallowed were plowed after harvest and left at rest through the following winter, then plowed again in early spring to destroy winter weeds and left at rest again during the summer, then planted in the fall. Although this practice resulted in acreage that lay dormant sometimes half the time, the increased yields and lower costs convinced farmers of its utility.

The wheat boom dramatically transformed much of California's interior landscape. The growing crop was a beautiful sight, its scale in those days epic, and its progress marked by splendid gradations of color. Lush broad fields of green sprouts in the winter phased into brilliant yellow-green as the stalks developed heads in the spring. Seasonal winds could create a waving ocean of grain in the wheat-growing areas, stretching for miles across open country, with the western vistas occasionally punctuated by cloud banks moving in across the coastal mountains to release showers visible from miles away. As summer approached, the wheat fields gradually turned bright gold, shining in the sun or rippling in the wind—then, if all went well, to crisp brown as the heavy-headed stalks waited for harvest.

From the coastal mountains to the Sierra, from the flat lands along the rivers to the rolling foothills, much of the Central Valley was converted into vast grain fields. "Wheat, wheat, wheat, and nothing but wheat, is what you see on your journey, as far as the eye can reach over the plain in every direction," observed Charles Nordhoff, a traveler through the San Joaquin Valley in the early 1870s. "Fields of two, three, and four thousand acres make but small farms."

A popular author invited by railroad promoters to visit California, Nordhoff wrote an informative book extolling the potential for farming here, focusing one chapter entirely on wheat production. Yet while he was an admirer of much that he saw and predicted a brilliant future for the state's developing agriculture, he did not admire some of what he observed of wheat growing in general, noting that much land was leased out by big landowners and that local farmsteads had few amenities: "The buildings put up for renters are...as slight and poor as can be....there is no vegetable garden, there are no trees, there is absolutely nothing to make life endurable or pleasant; and the only care of owner and tenant is to get as much wheat out of the land each year as they can at the least expense....It is not a pleasant system of agriculture, nor one which

NOVEMBER : PLOUGHING

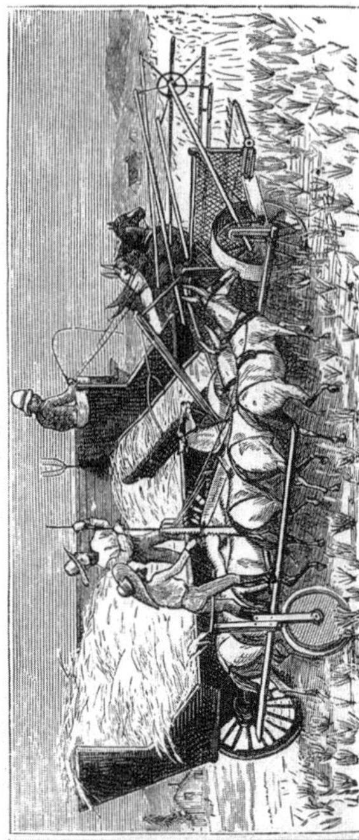

JUNE : HEADING

JANUARY : BROAD CAST SOWING

The stages of wheat production are captured in this 1883 engraving from a popular journal. *Courtesy of California Historical Society, North Baker Research Library, San Francisco.*

can be permanent; but it develops in the farmers who practice it a great deal of enterprise, and very shrewd business habits. They make money by economy in cultivation, and they are very quick in seizing new labor-saving devices. If I may judge from those I met, I should say that these wheat farmers are an unusually intelligent set of men, with great courage...." [2]

Hugh Glenn was one of those men. At this period of his life Glenn's hands were said to be "hardened and horny from actual labor," he had a "bronzed and sunburnt face" beneath red hair and behind a full, flowing beard, and wore plain but neat clothes. An active man, he could ride all day long on a horse or a mule but seldom engaged in recreation. He was said to be quiet and unobtrusive in manner and was regarded by at least some as a "good neighbor," [3] but he was also highly ambitious. He steadily acquired more land in the Sacramento Valley as it became available, eventually coming to control between 60,000 and 80,000 acres including tracts beyond his main holdings in Colusa County. The official map of Colusa County of 1874 shows his properties primarily west of the Sacramento River extending north from Princeton almost to St. John, where Stony Creek flows in to join the river. His contacts with Sacramento moneylenders helped finance his expansion, but he earned his reputation as a highly successful wheat grower through years of hard work and his own outstanding executive ability.

The Glenn family moved to the Jacinto Ranch in 1868, headquartered near Cid's (also spelled Sidd's) Landing, a steamboat stop 17 miles north of Colusa and four miles north of today's hamlet of Glenn, just south of Bayliss Road leading west from the river. The first Glenn home at Jacinto was a rambling one-and-a-half-story affair near the river, consisting of many additions to the original four-room structure built by Isaac Sparks and eventually containing some 20 rooms. Within a few years the Glenns had another, finer house built farther back from the river, a three-story structure with 22 rooms, some with marble fireplaces and elaborate woodwork. The old home was converted into a boarding house for employees, and the family began developing a California-style country estate with a lawn, flower gardens, and exotic trees like palms and eucalyptus. Glenn also planted 300 acres of vineyard as well as an orchard of fruit trees. Here Dr. Glenn—as he liked to be called—established an oasis in raw territory, living graciously and offering visitors southern hospitality in accordance with Missouri traditions, with good meals and fine whiskey.[4] As on many western ranches, there was a Chinese cook in the kitchen, duly noted as a member of the family in the 1870 census.

Nancy Glenn, a brunette with fine features, was said to be a kind and humorous friend and hostess, devoted to her children. Her ninth and last child was born at Jacinto in 1870. She suffered the death of her 10-year-old daughter Eva in 1874, but otherwise oversaw the active life of a growing family. The Glenn children had private tutors and were sent to boarding schools when old enough, but at the ranch they enjoyed boating, fishing, swimming, horseback riding, and the leisure activities of well-off families. There were plenty of other children around, too, for the ranch had its own school and teachers for the children of employees. This school eventually became part of the county school system, and by 1880 the Jacinto school district was reported to include a total of 66 children between 5 and 17 years of age (although only 38 were attending school), plus 31 children under the age of 5.[5]

Under Glenn's guidance the settlement at Jacinto grew rapidly. By 1872 the community housed some 350 employees and their families year-round, and numerous barns and corrals sheltered the hundreds of horses and mules needed for ranch work. Two merchants ran a brick store where ranch employees and neighbors could buy general merchandise, hardware, ranch supplies, and medicinal products (but no liquor). One corner of the store housed the office of Glenn's business manager and a small post office, for which Glenn served as official postmaster. The village also included two large warehouses and a creamery along the river bank as well as a ferry crossing over the river leading toward Marysville.

Jacinto served as a stagecoach stop for one of the regional stage lines operating out of the town of Colusa. The north-south stage line, usually carrying nine to fifteen passengers, transported mail for the upper valley towns along the bumpy river road, providing another transportation lifeline to the far reaches of the valley. Jacinto's two-story wooden frame hotel provided rooms upstairs mainly for the accommodation of tired passengers, but Glenn himself kept a bedroom there for convenience. Behind its open porch the hotel also included a popular saloon. The large stable building nearby kept swing teams ready to replace tired stage teams, and Glenn also maintained a regular stage service to Willows, the small community to the west.

Jacinto became an important wheat shipping depot. Boats arrived twice weekly during the shipping season to deliver supplies and take away grain and other products from the warehouses. There was no permanent dock at Sidd's Landing, since the riverbank tended to vary with the years, so boats were usually anchored to a big oak tree along the bank, with temporary plank walkways arranged from boat to shore. When the river was high enough, wheat was shipped by the thousands of sacks, loaded onto flat barges towed by small steamboats, each barge carrying about 100 tons. At season's peak the Jacinto trade might engage as many as seven steamers and 23 barges. The larger boats could tow five barges each, at a rate of

$1.50 a ton for wheat shipped to San Francisco Bay, where it was transferred into sailing ships bound for England. From Jacinto to the Bay ordinarily took three days.

Despite his growing commitment to wheat, Glenn never stopped being a livestock man. Having earned his living for years as a livestock trader, he took advantage of his spread-out acreage to maintain large numbers of animals. At the peak of his career at Jacinto he kept more than a thousand work horses and mules, hundreds of brood mares and young colts, and several thousand head of cattle, sheep, and hogs. He knew his animals. He kept excellent riding horses and was said to be very proud of two matched pure white mule teams. At least a quarter of his land was used to grow the barley and hay needed to feed the stock, and mountains of manure from the barns and corrals went into the fields to condition the soils.

Glenn had his eye on potential profits where he could find them, however, and as he converted his Jacinto ranch as rapidly as possible into wheat fields, he invested in livestock ventures elsewhere. In 1870 he bought a ranch in remote Washoe County in northwestern Nevada to raise stock on the range there in partnership with his nephew E. Waller Crutcher. Because of its isolation and limited resources, this ranch never made money, but in 1872, trying a different tack, he had one of his employees, young Peter French, take a herd of 1200 cattle with half a dozen vaqueros to the high desert of southeastern Oregon, just then gaining a reputation as prime livestock country. This decision proved prescient.

Peter French was born in 1849 in Missouri but moved as a child to California, where his father ran a successful sheep business for many years in Tehama County and his uncle farmed just west of the Glenn holdings in Colusa County. Peter grew up full of energy and totally at home with livestock. In his early 20s he took a job at the Jacinto Ranch, and Glenn soon recognized his talents. Peter proved his mettle in trailing the big herd to the relatively unknown territory in Oregon, and he became a highly competent and remarkably loyal ranch manager there for Glenn, who ultimately acquired some 150,000 acres in remote Harney County. The big "P Ranch" in the Blitzen Valley flourished under young French's management. Cradling the marshes along the Donner and Blitzen River and extending over an enormous rim-rocked plateau of rolling sagebrush, the ranch over the years produced many thousands of head of prime cattle for the California and Midwestern markets. Although there is no evidence that Glenn himself ever saw it, the P Ranch became one of the great historic spreads of southeastern Oregon.[6]

Wheat growers were gamblers. They had to be. Like many others in California they were speculators and risk-takers, borrowing money that they hoped would make more money and betting their crops against weather and markets. Those who knew something about farming in other regions had to re-learn much of what they knew, and their learning curves had to be sharp. Growers took chances on the fertility of soils that they hardly understood. Immigrants from the Midwest with its relatively narrow range of soil types were ill prepared for California's hugely variable, disparate soils, where friable loams could yield to thick adobe sometimes within a matter of yards.[7] The California State Agricultural Society, which issued annual reports on farm subjects, ran articles in its early years on fallowing and other soil conservation techniques, but probably influenced fewer farmers through its publications than through its exhibits at the state fair. With land-grant university research and agricultural experiment stations still well in the future, farmers in the 1870s had to learn from their own experience, and though many began keeping records on their own, it would take years before clear patterns would be established through systematic data collection by professionals.

Weather was always an unpredictable variable, especially in the years before local weather patterns were documented. "Rain is the staple of conversation in the country, next to the Liverpool market," reported one observer.[8] Wheat could suffer from the unreliability of rain—too little, too much, or not at the right intervals—and the possibility of occasional killing frosts or untimely hot winds. Annual temperatures in the Central Valley, though generally favorable toward year-round agriculture, could vary from below freezing in winter to well over 100 degrees for days at a time in summer. Because wheat growers depended on one crop alone, even normal variations, depending on timing, could bring success or disaster. In some areas of California wheat growers could count on only three good harvests in five years—gamblers' odds indeed.

Severe droughts again hit the state in 1870–71 and in 1876–77, crippling wheat production, causing the death of millions of sheep, and creating widespread distress especially throughout the San Joaquin Valley. On the other hand, flooding from extended winter storms could turn vulnerable fields into gooey mud or wash out crops altogether. In 1878 January and February floods brought the Sacramento River and its tributaries to their highest levels in years. At Jacinto Glenn was reported to have lost 2500 sheep and 200 hogs in the bottom lands during that period of high water, and a good portion of the valley's wheat crop that year was wiped out, while the towns of Willows and Colusa were islands in a sea of water for weeks.[9]

Other threats to the wheat crop included competitive weeds and diseases like smut and rust that could wipe out a long-awaited harvest. Not least were the farmer's age-old enemies, gophers and ground squirrels, which tunneled opportunistically through plowed fields and raised big families on a newly generous diet. Many farmers battled the resistant little beasts with traps, strychnine, or patient surveillance with shotguns. Then there were the rabbits: some communities began holding rabbit round-ups, in which hundreds of the marauding creatures might be pursued by dozens of men and teenagers across fields into fenced holding areas and systematically shot or clubbed to death. Early photographs of those events show proud victors standing behind mounds of dead rabbits, well satisfied with the at least temporary destruction of one of the threats to farm survival.

A major challenge to raising wheat in the Sacramento Valley was its location along the historic Pacific flyway for migratory waterfowl. Wild ducks and geese began arriving from the north by the millions from mid-September through December, at just about the time the ground was being prepared and seed sown for the winter. Settling in on the extensive marshes in the lower valley, as they had for centuries, huge flocks feasted through the winter months on the natural vegetation—or now the newly planted grain—until their spring migrations took them north again. Portions of the valley were called the "goose lands" for the enormous clouds of birds that rose at daylight and sailed in formations throughout the day, their squawks resounding for miles. So numerous in those days that their passage could cast shadows as the flocks flew over the fields, the geese would often drift down to eat the emerging crops. Farmers considered them worse than locusts, for unlike locusts they could do tremendous damage each and every year.

To protect their assets, farmers began to hire "goose-herders" to protect their grain fields during the growing season. Glenn's costs for herders, like his fields, were enormous. At a dollar a day for la-

Rabbit drives were community events in which dozens of farmers and their families collaborated in the massacre of one of the pests ravaging their crops.

Courtesy of Willows Museum.

bor, his goose-herding bill during the 1870s and 80s ran to $30,000 a year, a figure that fascinated journalists. It could cost as much as 50 cents an acre to protect the crops from depredation. Goose-herders in the fields worked in shifts, using Winchester shotguns from horseback to prevent the birds from settling. Cabin herders with dogs also actually lived in the fields, each one responsible for about 160 acres for $30 a month plus board. These men could also sell the feathers from the birds they shot and plucked, earning more money. Although Glenn hired goose herders up until his death, other valley farmers in the 1880s began contracting with market hunters, whose improved shotguns, it was claimed, could take down 60 to 100 birds at a single shot. Market hunters could sell their kill in San Francisco at about $1 for a dozen ducks or geese.[10] Millions of birds thus vanished from the Pacific Flyway over the space of only a few decades–the California equivalent of the vanishing buffalo on the Great Plains.

Shipping was a gamble too. Shipping costs from California ports were higher than from other U.S. ports because the distance to Europe was much longer. Because transcontinental rail freight charges were prohibitively expensive, most California wheat exports moved by ship out onto the Pacific Ocean and down the coast of South America to perilous Cape Horn, then northeast across the Atlantic Ocean to the British isles–a journey of more than 14,000 miles that crossed the equator twice. From San Francisco to Liverpool, the main destination, the voyage lasted from 100 to 130 days or an average of about four months. It was an immense undertaking.

Getting the crop out from the interior valleys alone was a prodigious effort. At first, before local railroads opened up inland transportation corridors, grain was an almost exclusively waterborne commodity brought out on riverboats to connect with world trade through the wharves of San Francisco Bay. During the 1860s Maine Prairie in the lower Sacramento Valley was the biggest interior grain shipping point in California, but soon other ports farther up the rivers loaded cargo from deep within the interior. In the 1870s grain warehouses became common along the waterways, and schooner scows and steamboats pulled barges mounded with pyramids of fat grain sacks slowly down to the great Delta where the rivers mingle, headed for the deep water where the sailing ships moored.

The deep-water vessels that took the wheat through the treacherous seas around Cape Horn and across the broad Atlantic included large English iron clipper ships and Yankee "Down Easters," medium clippers built in Maine. Braving fog and storms, currents and rocky coasts, these sailing ships took precedence over steamships because the shipping costs were less. A maritime historian writes: "The development of the Pacific coast...opened a new traffic for the noble, white-winged three-masters. The grain of the Pacific slope could not bear the railway charge to the East, but it could bear the ocean rate around Cape Horn. To San Francisco and Portland now flocked the largest and best sailing vessels of all the maritime nations of the world. Grain was a treacherous cargo, liable to shift to one side, if loaded in bulk, and always imposing a severe strain on the hull that conveyed it. Only the strongest and most powerful ships and the boldest and most experienced seamen were fit for the fourteen-thousand-mile voyage around the stormy Cape." [11]

In the famous wheat fleet that plied the oceans, American shippers handled only about a third of the total trade. "It was a contest of truly Olympian dignity–of the best ships of many flags with each other and with the elements. Out through the Golden Gate there rode every year in the later seventies and the eighties, southward bound, the long, lean iron models of Liverpool and Glasgow, the broader-waisted, wooden New Englanders, with their fine Yankee sheer and tall, gleaming skysails, the sturdy, careful Norwegian and German ships, often launched on the Penobscot or Kennebec, and here and there a graceful Frenchman or Italian. The British were the most numerous, because the total tonnage of their merchant marine was by far the greatest. Next came the Americans." [12]

After the Civil War American deep-sea maritime traffic actually went into decline, and the bulk of the California grain trade was ultimately handled by Great Britain. Although American ships were said to be more efficiently managed and faster by days to their destinations than their British counterparts, they were subject to discriminatory insurance rates by Lloyd's of London, which discouraged American shipping.[13] The California wheat trade became an example of the long imperial reach of Victorian England's mercantile interests, dominated by British financial institutions with their connections to the Liverpool Corn Exchange. Two British banks were in fact among the leading institutions in San Francisco by the 1870s, and the British investment house of Balfour and Guthrie was for years a principal broker for the California wheat trade. Incoming cargo on chartered ships was often British coal, cement, manufactured goods, or the always popular alcoholic beverages, and wheat was the commodity that paid the way home.

Prices for California wheat growers were ultimately set in faraway places like Liverpool or Marseilles, for the grain markets along the East Coast were usually supplied by the Midwest. Growers and the merchants who depended on them watched the daily fluctuations of the Liverpool market with anxious solicitude. Because California wheat was hard, dry, and unusually white, it commanded premium prices in the international markets despite the enormous distances and time involved in shipping it.

Here again, growers took a gamble. Access to market information by the new transcontinental telegraph and transoceanic cable helped with chartering oceanic vessels and reduced the risks of grain dealers between harvest and sales. But generally it was the dealers who had that access, not the growers. The urban middlemen who controlled marketing had farmers in the spread-out rural counties of California at their mercy, and brokers took their cut at every step of the transfer from producer to consumer. Control of grain sack production, storage, and transportation was key to control of the entire trade, and individual farmers might well come out on the short end of transactions. Manipulation of crop prices and imposition of heavy service charges could eat up any profits they might hope for. Even during their good years, wheat growers knew that their brokers were doing better than they were.

This was not a problem for California alone. During these years Midwestern grain barons like Pillsbury and Cargill made fortunes from handling the raw commodities, and across the country farmers continued to struggle against the power of elevator and mill owners, railroads, and shippers to determine their earnings. Though the Chicago Board of Trade was established in 1848 to help bring order into a chaotic selling scene, for years thereafter the exploitation of the nation's grain trade remained a simmering scandal across America. Before effective controls were put in place, "cornering" the trade (buying up all stock options) made fortunes for some while others went broke.[14]

As in Chicago, traders in the California wheat market tried to run corners or to monopolize shipping, including such well-known capitalists as James Flood, James Fair, George McNear and others. None, however, were as successful or recognized as Isaac Friedlander. The career of this flamboyant character reflected the ups and downs of an inveterate gambler.[15]

After his early speculations in flour and flour milling, Friedlander soon realized that domination of shipping was central to the fate of the entire California wheat crop. Vessels had to be chartered well ahead of time, for many ships would not call along the Pacific Coast unless they had both a ready market for their incoming goods and a guaranteed cargo to take back out. Friedlander began a system of estimating each year's crop before it was harvested, relying on a network of local agents like William Dresbach in the lower Sacramento Valley to report on weather and growing conditions. He would then charter vessels in the spring before other brokers began negotiating. He would also buy quantities of wheat on contract and resell it to export houses, providing them also, of course, with the shipping to move the grain abroad. At a two to three percent commission on his transactions, he did very well indeed. He was bold and quick, and his judgment was usually sound.

The genial Friedlander became very influential, for he was widely regarded as a far-sighted and clever businessman, and his lavish hospitality in his city mansion helped him make important friends. He was elected one of the first presidents of the San Francisco Chamber of Commerce and was appointed to the first Board of Regents for the young University of California. Always attracted to moneymaking schemes, in the early 1870s he moved into new ventures in land and water development. He joined his friends the banker William Ralston and the capitalist William Chapman in buying big chunks of newly opened land in the San Joaquin Valley and in planning to provide irrigation for hundreds of thousands of acres of that land by building the San Joaquin and Kings River Canal Company—a story told in a later chapter.

Isaac Friedlander stood out in San Francisco as an unusually energetic businessman, a grain trader in the world market and an investor in Central Valley land development. *Courtesy of California Historical Society, North Baker Research Library, San Francisco.*

By 1872 Friedlander dominated most of California's grain shipping, and a ship was even named for him. At this point, however, the farmers of California rose up in revolt, in a brave but short-lived effort to take marketing into their own hands.

The Order of the Patrons of Husbandry was founded in 1867 in Washington D.C. as a fraternal organization for farmers. By the early 1870s many farmers in the Midwest had joined the new association in order to fight the power of the railroads and grain elevators over their livelihoods. Popularly known as the "Grange" for its community halls in rural communities, the organization had several purposes but quickly became best known for its anti-monopoly activities. Grangers' struggles in the Midwest were mirrored by the "great California grain war," as described by historian Rodman Paul.[16] Like their brothers elsewhere, farmers in California resented the grain merchants who manipulated them and tried to fight back.

California growers were not interested in Chicago grain prices or in railroad freight charges, but in Liverpool prices and oceanic shipping rates. At this point the Central Pacific was not yet entrenched enough to monopolize all inland transportation, so most farmers were still using riverboats. They were also still using sacks for their grain, rather than handling it in bulk, for sacks made for convenient stacking on barges and wharves and for safer shipping by sea, even though handling the grain this way was more expensive. Thus, unlike the Midwest, there were no elevators in California (Isaac Friedlander had built one as an experiment at Vallejo in 1869, but it was highly controversial and then collapsed in 1872). San Francisco brokers speculated in sacks as well as in ship charters, raking in profits from a product they bought cheap wholesale and sold retail to individual farmers.

Scattered in rural areas without convenient means of communication, early farmers lacked the ability to resist the exploitation until they could somehow organize themselves. As early as 1870 farmers in several counties had recognized the need to band together to promote their own education and self-protection, and had begun to establish local farmers' clubs. After the *Pacific Rural Press* began publication on January 7, 1871, this journal became a valuable resource in many farm homes and a leading advocate for farm organizations, reporting on the activities of many rural associations.

The Napa County Farmers' Club was the first to address the problem of middlemen. This group made a proposal for farmers to charter their own ships for exporting wheat, crafting a resolution in early 1872 to organize for this purpose statewide and even citing the infamous Friedlander by name as the controller of some three quarters of the wheat trade. A few months later several local clubs joined together to form the California Farmers Union, electing John Bidwell as its first president. Split between the interests of big and small farmers, however, this larger organization grew only slowly. Its first effort to negotiate lower sack prices ended in failure when city sack dealers formed a combination or "ring" and then actually raised prices. That fiasco effectively spelled the end of the Farmers Union.

The Napa club then discussed forming a state branch of the national Order of Patrons of Husbandry, and in March 1873 the Napa farmers elected to reorganize their club as a local Grange. This marked the entry of the national organization into California. The movement grew rapidly, encouraged by reports of successful activities in other states. By July thirty-three local grange associations had been formed, and a State Grange was organized. By 1874 California had 231 local granges with a membership of more than 13,000, including women. This was no peasant uprising, however, for most members were relatively well educated and prosperous family farmers, and they intended to improve their circumstances through organized political and economic action.

Despite all good intentions, the drama of the Grange revolt in California was brief and highly disappointing. In the summer of 1873 the Grange sent a committee to San Francisco to negotiate shipping with the firm E. E. Morgan's Sons. The firm was asked to charter ships for the use of wheat growers, who would then export their crops under their own shipping accounts. Morgan's agreed to do so, but that year turned out to be a short crop year, both abroad and in California, and some farmers decided to hold back their harvested wheat in anticipation of higher prices. When the chartered ships came in for loading, Morgan's was forced to buy wheat to fill them and went into debt. By the end of that year the economic panic that had started earlier on the east coast hit San Francisco, further roiling matters. On Grange request Morgan's had also ordered two million sacks for the harvest, but the sack "ring" in San Francisco now strategically cut their prices, and individualistic farmers again bolted the Grange agreement and bought cheap. Morgan's lost again. Although the next year turned out to be a good crop year, there was a bear market on shipping in 1874. To beat the competition Morgan's had booked ships early and already paid relatively high charter rates. The firm went bankrupt. Farmers had to sue to collect any payment whatever for their wheat, and the Grange venture ended in dispiriting failure.[17]

Having waited out the debacle he saw developing, Isaac Friedlander went back to his usual grain trading. But in 1877 he too declared bankruptcy. His investment in the San Joaquin and Kings River Canal Company had ended in financial disarray, and when his friend and associate William Ralston of the Bank of California died unexpectedly in 1875, Friedlander lost some of his easy access to credit. An incurable speculator, he too over-chartered vessels to export a wheat crop that turned out to be smaller than he had estimated, and lost another fortune. Though he began trying for another comeback, he died the next year at age 55, weighing 300 pounds, of apoplexy. His estate was so encumbered by debt that it took some years to pay off the creditors.

---

1 Sources for 1870s wheat prices include Fraser, "Rainfall and Wheat in California," and Green, *The History of Colusa County.*

2 Nordhoff, *California for Travellers and Settlers*, pp. 182, 187. Nordhoff's commentary on the San Joaquin was usually more favorable.

3 Quotes from a story on Glenn in the *Chicago Tribune*, July 31, 1879.

4 Sources on Glenn at Jacinto include several contemporary newspaper stories; White, *Glenn County Sketchbook;* Allen, *Hugh Glenn;* and McGowan, pp. 258–60.

5 Green, *History of Colusa County.*

6 Sources for P Ranch history include French, *Cattle Country of Peter French;* Gray, *Life and Death of Oregon Cattle King Peter French.*

7 The official mapping of California soils took decades, beginning with government surveyors in the 1850s and continuing with soil scientists in the College of Agriculture at UC Berkeley until well into the 1900s.

8 Davis, "Wheat in California," *Overland Monthly,* 1868.

9 Weather of the 1870s is described in Green, *History of Colusa County*; McGowan, *History of the Sacramento Valley.*

10 McGowan, pp. 363–75.

11 Marvin, *The American Merchant Marine*, p. 383ff.

12 Marvin, ibid.

13 Marvin, ibid. For discussion of the British role in wheat shipping, see also Canright, "Outpost of the Empire: The British Domination of the California Grain Trade."

14 Morgan, *Merchants of Grain.* This history of the international grain trade includes two chapters relevant to the California wheat boom: "Bread and Dynasties" and "Grain Barons."

15 Biographical information about Isaac Friedlander has been pieced together from *Merchants of Grain*, Rothstein's "The Wheat Kings," and folders at the California Historical Society Library in San Francisco.

16 Paul, "The Great California Grain War"; see also his "Wheat Trade between California and the United Kingdom, 1854–1900."

17 Vaught covers local effects of the Grange effort in *After the Gold Rush.*

~5~

The Wheat King

DEVELOPING AND RUNNING A LARGE WHEAT FARM took a high degree of executive talent. Hugh Glenn was not a typical farmer, but he learned to understand the farming business intimately, from field cultivation through international marketing. As his holdings grew, Glenn had to manage hundreds of employees and the thousands of animals used in farming operations. He had to plan his farm schedule a year or more in advance; to seek and manage large-scale financing; to order and distribute tons of supplies; to implement good farming practices adapted to differing field conditions; and to keep abreast of new developments in technology and machinery. He had to learn the properties of varying soils, the ways water flowed in the sloughs and across the land during the rainy season, the impacts of wind or temperature on the growing crop. He had to be efficient or he would go broke: even the procurement of sacks for shipping the crop might account for five percent of the value of a harvest. And he had to sell his wheat for some kind of profit, whatever the vagaries of the market or the commissions of the middlemen. During the height of his career Glenn installed a 27-mile long private telegraph line from Colusa and employed an operator to keep in daily touch with grain buyers. Later he put in his own long distance telephone line to keep abreast of prices through an office in San Francisco.[1]

The Glenn holdings in the Sacramento Valley were divided into nine main units or sub-ranches, several of them leased by tenant-operators, each with a house and barns, blacksmith shop, and other farm-related outbuildings. This worked out well for both Glenn, who had responsible partners to help in the overall development and supervision of the ranches, and the tenants themselves, who hoped through the partnership to move a step closer toward ownerships of their own. In 1874 the lessees producing wheat and forage crops included Isaac Bayliss and George V. Hoag, the latter not only a farmer but an accomplished mechanic and engineer.

The size of Glenn's entire enterprise was enormous, with one field alone said to encompass more than 10,000 acres. Dozens of eight-mule teams were kept on the ranches to perform the hard work of cultivation and hauling. Glenn used them also to make improvements to the land as it was developed for grain. He dammed up sloughs and built levees to control water and minimize flooding, dug water holes for livestock, and installed 150 miles of fencing to keep stray animals out of the fields. He learned how to manage even the heavy soils back from the river, formerly thought to be almost useless for agriculture, so that they too produced grain crops. Every year was a learning experience, and he learned.

Bonanza farms like Glenn's were characterized by a growing emphasis on efficiency and the steady evolution of technology—hallmarks of the industrialization of farming beginning across America. Big farmers tried to take advantage of new developments in the rising profession of engineering. New kinds of farm machinery could cut down on labor needs and make it possible to move crops more rapidly to market, in the long run increasing agricultural productivity by making it less expensive.[2]

The wheat ranches of the Mexican era had used simple single-blade plows and brush harrows to plant and cultivate the crop, and hand scythes to harvest it. The wheat had been threshed by driving horses across piles of stalks laid in corrals, and then winnowed from the chaff by Indian workers tossing the grain in baskets on windy days–inexpensive but wasteful methods, dependent on many laborers. Post-Gold Rush farmers quickly improved on these primitive methods and began trying to copy the farm machinery then being introduced in the Midwestern states. Gang plows were introduced into California about 1860, consisting of sets of six to eight curved iron plowshares drawn by eight or more draft animals. These devices greatly increased overall plowing speed and effectiveness in breaking up soil. Hand seeding gave way to wheeled broadcast seeders by the late 1860s. Improved as these methods were, at the big Glenn holdings each section still took an eight-mule team a whole day to make a round with a gang plow, followed by the seed sowing wagons covering a 30-foot-wide strip and then a harrow team to cover the seed.

New technology in the fields made the most dramatic difference during the harvest season, which had always required the most manpower. Cyrus McCormick, a mechanical wizard who put together his first crude metal reaper at age 22 in 1831, developed a solid business building innovative farm machinery in Chicago during the 1840s. In 1847, just after the repeal of England's Corn Laws and a year before the opening of the Chicago Board of Trade, he opened a factory to produce his new harvesting machines. McCormick soon emerged as a central figure in the rapidly changing farm world. He became internationally famous after winning the grand prize for his mechanical reaper at the 1851 World's Fair in

London. When the Gold Rush and then the Civil War made American farm laborers scarce, the idea of mechanical harvesting attracted those farmers who could afford it, and McCormick's reapers sold very well during the 1860s.[3]

Glenn, who kept apprised of new developments in the Midwest, purchased some of the McCormick reapers available after the war. He even wrote to McCormick personally to tell him that his machines were too small for California, inviting him to come and spend time on the Jacinto ranch to design equipment more appropriate for western conditions. McCormick accepted this invitation, and the two men in consultation came up with a new model. One of the enterprising Studebaker brothers also visited at Jacinto, and Glenn later ordered wagons from that company adapted to his own suggestions.

Pacific Rural Press readers read the national reports and advertisements for mowers and reapers that were carried in that journal, but the distance to the Far West from Chicago factories made most machines prohibitively expensive to ship. Many California farmers were themselves good mechanics, however, and some were expert blacksmiths. In the 1860s several began building their own farm implements on the ranches where they were to be used. A Marysville factory began turning out four-share plows for sale, and other machinery manufacturers started commercial production in Stockton and Oroville. By the late 1860s at least three inventors were also conducting demonstrations of steam plows that drew widespread attention. Thousands of people viewed one of these machines at the California State Fair of 1868, and many farmers went home with dreams in their heads.

The large-scale mechanization of wheat harvesting was particularly successful in the Central Valley because of its mainly flat terrain, large fields, and very long rows of grain. Seasonally dry soils at harvest time allowed for the use of heavy machinery that might bog down in rainier regions, and the flinty kernels of wheat could withstand the metal parts that touched them. The general scarcity of reliable labor in spread-out California also encouraged farmers' interest in mechanical harvesters as they became available. By the late 1870s teams of eight to twenty horses or mules were pulling large mechanical reapers in the major wheat areas, cutting swaths up to 40 feet wide, and this kind of horsepower replaced centuries of old-style manpower.

The wheat harvest was the most dramatic part of the year. Beginning in June, the harvest could last through most of the summer on the big ranches where the fields were planted in sequence. At first the usual practice was to send out headers pushed by mule teams—machines that cut the stalks slightly below the heads of grain—followed by gathering wagons that took the grain to a bulky stationary thresher-separator run by a steam engine. The thresher-separator, looming over the field, was positioned after clearing a large circle in the center of the section. Straw was used to fuel the boilers despite the ever-present danger of fire (in some areas prudent farmers took out insurance for up to half their crop). As the grain emerged from the separator, workers rapidly filled sacks, sewed them up, and threw them onto the stubble to be gathered by wagon teams taking them to storage.

A description of the California wheat harvest in a popular journal reveals one observer's mixed feelings about the mechanization of the process: "Alas for the romance of the harvest; the sickle, the cradle and the flail, the reapers and the gleaners—Boaz and Ruth—all are gone! The picture now is a broad, hazy plain, bounded by brown hills, which flicker and glimmer in the mirage; no trees, no running brooks, no green grass, but miles on miles of grain. Far away you descry clouds of yellow dust, and as you come nearer, you see the wagons drawn by horses coming in loaded with piles of grain and returning empty, and in the centre stands the huge machine, driven perhaps by steam, perhaps by a score of horses traveling in an endless circle, and fed by men dark as mulattoes with the sun and dust, perhaps with mouths and nostrils swathed to protect the lungs from the dust. There is no romance in this." [4]

One of Glenn's tenants, the talented George Hoag, put together a huge grain separator for Glenn that he named the Monitor. Combining three smaller separators, the Monitor could thresh up to 1,700 sacks daily or about 3,825 bushels. On August 8, 1874, the Monitor set a threshing record of 5,779 bushels in a single day, an enormous output that became famous in the farm world and the source of great pride for both Hoag and Glenn. Hoag was so proud of his invention that he commissioned the well-known Bay Area painter Andrew Hill to paint it. Hill visited the Glenn Ranch to create "A California Harvest Scene," a large oil painting depicting dozens of men, animals, and wagons bringing grain to a bulky machine. The *Pacific Rural Press* carried an engraving of this painting on its May 6, 1876, front page along with a story noting the "small army" of workers employed by Dr. Glenn, the "farmer prince." [5]

In 1877 Glenn ordered a bigger engine for the Monitor, and Hoag built an enormous straw burner steam engine that he called the Missouri Chief. The Monitor then broke its own 1874 record in July 1879 with 2,748 sacks or 6,183 bushels cut and threshed by a crew of more than 100 men running ten headers and 36 header wagons. In the circle of farm machinery enthusiasts, many were awed. Nevertheless, the huge steam engine was dangerous: a year later the firebox of the Missouri Chief blew up and killed three men, badly injuring several others, in a huge explosion heard from miles away.

Disasters like this were not the only hardships for farm laborers of the time.

Hundreds of men worked for Hugh Glenn as his wheat ranch grew. By the late 1870s he employed as many as 600 during the peak months of harvest, half that number during the winter. During the harvest season all employees and Glenn himself worked from dawn until dusk, 12- to 14-hour days with little break. On the government land just west of Jacinto lived a number of squatters who worked seasonally for Glenn, but mainly the harvest crews were "bindlemen" or "bindlestiffs," itinerant workers carrying their own bedrolls or "bindles." These men appeared when they heard there were jobs available and disappeared when the harvest was in—or sooner, if they felt like it. Most of them were young white single men, Americans or European immigrants, and some of the faces in the old threshing photos look very young indeed. Some were Irish immigrants who worked on Bay Area docks in the winter or found odd jobs in the cities. Some moved between seasonal jobs in logging, mining, or other resource-related occupations, while others were drifters or what the *Chicago Tribune* dismissively called "floating class" workers who did not want permanent jobs even when offered.

The going rate for common harvest labor during the 1870s was $1.00 to $1.50 for a 12-hour day, while a good sack sewer on the separator might make as much as $2.50. A sack sewer had to be something of an athlete, strong, quick, and well-coordinated, for a sack weighed 100 pounds or more, and the sewer had to fill and truss each one as the grain continuously flowed out of the separator, then toss it down from the machine for pickup by other workers—no mean feat. Drivers of field machinery also earned more, for they were responsible for making field repairs when something went wrong with the belts, pulleys, gears, or wheels of the complicated equipment, and they had to be good and even ingenious mechanics as well as competent drivers of bulky vehicles. Other skilled workers—blacksmiths, carpenters, and farriers—could earn $50 to 60 a month.

All of the harvest work was hard, made more exhausting by sometimes severe heat as well as dust and chaff, black gnats and mosquitoes. Even in June Sacramento Valley temperatures could soar well above 100 degrees, threatening the unwary with heatstroke. Field workers tried to shield their faces with bandannas against the thick dust thrown up by the machinery and the clouds of biting insects. And the work was potentially dangerous: accidents were not uncommon as tired laborers bumped into machinery, perhaps losing fingers, hands, or even arms in gears that lacked shielding. Men could be maimed or killed by steam engine explosions or even by cantankerous mules. They learned to watch out for rattlesnakes between stacks of grain sacks or parked wagons.[6]

Itinerant laboring men led a hard, coarse life up and down the valley, carrying a blanket roll from job to job as they checked into ranch bunkhouses. They came from a wide variety of backgrounds and worked for a variety of purposes. Some were fairly well educated, though most were not, and some saved their money carefully for some financial goal, though most did not. Most did not care to share their pasts with others, and few asked. There was always a certain criminal element, men who kept their heads down to avoid punishment for infractions committed elsewhere. Violence was not uncommon. In one fifteen-year period Colusa County authorities recorded 69 murders, twice that number attempted, and as many suicides and other violent deaths in a population of only 15,000, many of whom worked as farm laborers. The infamous Dalton Brothers, wanted for a series of robberies in the plains states, claimed to have worked on the Glenn ranch before they were apprehended for holding up a Southern Pacific train in the San Joaquin Valley in 1891.

Harvest jobs included board and room besides the modest wage, but the lodging was always minimal. Cramped bunkhouses, built of rough boards, usually had no window screens to keep out the ubiquitous flies and mosquitoes. In summer they were hot and smelly places, in cooler weather cold and drafty. The food was better than the lodging because it had to be. Hard physical labor made for huge caloric needs, and employers knew they would lose workers if they were not well fed. A water wagon and a cookhouse followed the men in the fields, and lunch break lasted up to an hour. The cooks, usually Chinese, prepared plentiful food, with beef, pork, mutton, and beans serving as mealtime staples, accompanied by vegetables and fruit as available, some kind of sweet dessert, and lots of hot coffee.

Though their workdays were arduous, field hands had occasional off hours and leisure time on weekends, and they found ways to entertain themselves. Some diverted themselves with card games, others teased companions with practical jokes, and a talented few played the banjo or told stories to entertain the company. Recreational wrestling was popular, but there were also serious quarrels and fights that might leave men bloody. A dance hall on the upper floor of the Jacinto Hotel brought men and women together for occasional socials on Saturday evenings. On other Saturdays many workers went off to the town of Willows to abandon themselves to whiskey, gambling, and easy women. In those days there were said to be 25 saloons and dozens of whores plying their trade along the town's main street. On Sunday afternoons the Jacinto foremen drove wagons to town to collect the men incapacitated from drink or brawling. The journeys back to the ranch were always quieter than those going in.

Chinese workers were a significant part of the labor force through the 1870s. They crossed the Pacific Ocean by the tens of thousands from the 1850s onward, attracted by stories of gold strikes in a far-off land. An estimated 10,000 Chinese arrived in San Francisco before 1852, and though they spread surprisingly quickly across the western states, most remained in California. Without their backbreaking labor the Central Pacific Railroad might never have been completed. By 1876 there were some 122,000 Chinese in California, nearly all men.

The Chinese were known as excellent workers, patient and thorough, hardworking and abstemious. They did work that white men refused to do. They sifted through mine tailings, laid railroad beds, built irrigation and reclamation systems, constructed levees, cleared land, cut fuel wood, and put up fences and stone walls. They made and sold brooms, dried and chopped tules for packaging, managed cook wagons, did general chores, mended sacks, and worked on harvesting crews, although not usually on the big machines. The "China boys" did nearly all the domestic work on California ranches, serving as cooks, cleaning men, laundrymen, and gardeners. Some Chinese grew market vegetables on leased or vacant land and peddled their produce with horse carts, selling to the big ranches and in town, and they fished in the rivers or picked wild berries despite the mosquitoes. The Chinese were steady and dependable, often ingenious in their solutions to common problems, and while they almost always earned less than their white counterparts, they boarded themselves and for the most part stayed out of trouble.[7]

Charles Nordhoff expressed considerable admiration for the Chinese, describing them as "very good men...civil, obliging, and competent." But he noted that their toil and efficiency did not endear them to some Americans, particularly the "Pikes," the migratory southern poor whites who were settling into various areas of the state and believed that their opportunities were diminished by Chinese competition. He wrote that "the Pike is the Chinaman's enemy. He does little work himself, and naturally he hates the patient industry of the Chinese. Of course, if you ask him, he tells you that he is 'ruined by Chinese cheap labor.' 'You could no more get these fellows to work than you could get grasshoppers out of a vineyard,' said a farmer to me; 'but they sit in the saloons, and growl about the unfair competition of the Chinese. One Chinaman is worth a regiment of them.'" [8]

The resentment of white working-class men began to grow into the collective prejudice of the majority during the lean years of the early 1870s. The virulent anti-Chinese feeling of the decade also grew out of racial stereotypes. Although "John Chinaman" as an individual earned solid respect from those who recognized his virtues, the Chinese as a group seemed clannish and strange to the average American. They spoke little English, kept to themselves, practiced strange customs, lived in overcrowded conditions, seemed to gamble incessantly and were known to smoke opium, and sent their wage money back to China. Even those who admired their work habits believed that they needed to become better integrated into society, especially in urban areas. Nordhoff called San Francisco's squalid Chinatown "a blot on the city," and the dirty, crowded Chinese sections of other towns were thought to be breeding places for vice and disease.

Two big Chinese communities grew up near the Glenn ranch at Jacinto. The Chinatown in the town of Willows to the west had several hundred workers crammed into a two-block area full of stores, laundries, card rooms, whorehouses, and opium dens. The village of Princeton along the river to the south had a smaller Chinese population. For the most part these workers–farm laborers, gardeners, cooks and laundrymen–led modest and industrious lives, punctuated by occasional more colorful moments at times of celebration or sorrow. Chinese New Year was always a source of fascination for onlookers, but other Chinese customs tended to puzzle the white majority. (The explosion of the Missouri Chief steam engine in 1880, for example, killed three young Chinese men who were pitching grain onto the separator, scattering their bodies and belongings for a hundred yards. The surviving Chinese covered the road leading from the wreck to Princeton with strips of red and yellow paper, to prevent the devil from getting the souls of the dead, they said; and the story was passed down by bemused locals.)

But the day of the Chinese on California ranches began to wane. The collapse of the stock market and many business failures after 1873 helped escalate prejudice against Asians, and Chinese laborers became a popular scapegoat. Their presence became a hot topic in some gatherings, and by 1875 some newspaper editors were openly arguing against continuing immigration. The *Gilroy Advocate* complained that "the whole land will be overrun with these Mongolians." Though farmers who employed them found them indispensable, workers in cities and towns grew more resentful. Hooligans attacked Chinese laborers for dubious reasons, sometimes beating them up. In San Francisco the Workingmen's Party implicitly encouraged violence, shrilling "The Chinese must go" at rallies calling for their deportation. In the Sacramento Valley several communities formed anti-Chinese societies. Feelings ran especially high around Chico, where there were so many attacks on Chinese that John Bidwell, a target of criticism for hiring them, organized a Committee for Safety to protect local farms and ranches from being threatened by thugs.

Anti-Chinese feeling peaked during the 1879 state election, and in response to the rising popular sentiment the U.S. Congress opted to fight the so-called "Yellow Peril" by passing the 1882 Chinese Exclusion Act (which was extended in 1892 and again in 1902). With new immigration effectively cut off, aging Chinese workers had few replacements, and the number of Chinese in rural areas gradually began to dwindle. In 1887 the Willows Chinatown accidentally burned down to the cellars and was never fully rebuilt. The dead were shipped back to China.

Some Sacramento Valley farmers in the 1870s were threatened by a fast-rising problem: the destructive effects of hydraulic mining on downstream land and water. Over the course of about twenty years, hydraulic mining in the mountains seriously damaged both river shipping and agricultural productivity in northern California, wreaking havoc on portions of the foothills and lower valley.[9]

The infamous 1862 floods began the problem, carrying so much mining sludge out of the mountains that year that some early farmers along the Yuba and Bear rivers simply gave up trying to farm and moved elsewhere. In normal or dry years the damage was less, though it continued, but in January 1875 a series of big storms again deluged the state. Marysville, devastated with flooding from the Yuba River, became a virtual dump for mining debris, and it took months to dig the city out of the mud. Afterwards Marysville built big levees, essentially turning itself into a walled fortress, while farmers in outlying areas began mobilizing for legislative action to help defend unprotected land against future catastrophes. During the winter of 1875–76 the struggle between mining and agricultural interests moved full force into the Legislature. The first proposal for a state investigating commission, however, narrowly failed the Assembly after mining advocates torpedoed the effort.

For the next eight years the controversy simmered, moving in and out of the Legislature and the courts. In 1878 a coalition of farmers formed an Anti-Debris Association to counter the influence of the Hydraulic Miners Association. In 1880 State Engineer William Hammond Hall reported that more than 36 million cubic yards of mining debris had been deposited in the watersheds of the Feather, Yuba, Bear, and American rivers, rendering the natural periodic flood hazards much worse. The bed of the Sacramento River near the city of Sacramento was reported to have risen an average of 15 feet between 1854 and 1879, and navigation along the entire river system had become more problematic with the years. Overflow debris had also buried some good farm land along the lower portions of the watershed under ten or more feet of sand, gravel, and "slickins."

The worst of these problems lay to the east and south of Jacinto, but Hugh Glenn met with other concerned citizens in Colusa County to send anti-mining-debris resolutions to the Legislature, urging action against the ongoing environmental disaster. He did not live to see the judicial injunction against hydraulic mining that was finally published in the 225-page Sawyer decision of 1884.

Though changes in river bed levels increasingly complicated shipping along the Sacramento River, steamboat traffic continued for a couple of decades more. At least a third of the state's wheat crop continued to move by water, for the freight rates by river were only half those of the railroad even after the new State Railroad Commission began to regulate rail charges. Glenn shipped his wheat both ways, sometimes down river by barge from Jacinto and sometimes out of Willows by rail. After February 1881, when very high water from winter storms damaged extended stretches along the river banks, Glenn and his fellow wheat growers increasingly turned to the relative reliability of rail shipments compared with the cheaper but less certain option of water transport.

The 1870 Census counted more than half a million (557,375) residents in California, up from 380,016 in 1860, but the real growth was yet to come. With the transcontinental railroad completed, California experienced a new surge of immigration nearly equivalent to that of the Gold Rush. Many of the newcomers hoped to acquire land, but would be disappointed.

One of the factors discouraging earlier immigration to the state had been the uncertainty of acquiring clear land titles because of the extensive but ill-defined Mexican land grants. Now, following the settlement of most of the land grant claims and the opening of large

Even after the railroad offered an alternative, river shipping was the preferred option for many grain farmers in the Sacramento Valley. This 1890 photo shows sacks being loaded in Colusa County on a typical small steamboat.
Courtesy of California Historical Society, North Baker Research Library, San Francisco.

areas of public land, speculation became the problem. Sophisticated investors with financial backing in San Francisco were able to manipulate the confused and poorly administered public land laws of the time to buy up huge tracts at cheap prices, tying up hundreds of thousands of acres with the intent of eventually subdividing them for large profits. This problem was particularly acute in the San Joaquin Valley, newly opened to public notice by the nation-wide promotion of the Southern Pacific Railroad. That company also hoped to cash in on escalating values in its checkerboard land grants along rail lines.

The grip of these early land monopolists deterred orderly settlement by small- to medium-sized farmers and created great controversy in the 1870s. One newcomer who allied himself with the capitalists was Theodore Kearney, a young Irishman from Boston with modest means and a shadowy past. Kearney stepped off one of the first transcontinental trains to arrive in California in 1869 and soon began working with San Francisco financier William Chapman to develop fruit colonies on land purchased in Fresno County. Some of these colonies became successful while others floundered, but Kearney himself amassed a fortune. Other prospective settlers without bankrolls or influential connections became frustrated. Squatting on little-used public or railroad land became fairly common, but the squatters' resentment against the big landowners did not contribute to social tranquility.

Henry George, a zealous young San Francisco journalist, took on the issue of land monopoly in his popular book *Progress and Poverty,* published in 1879. In this early work of economics based on observations of California in the 1870s, George argued that land speculation led inevitably to the unjust distribution of wealth, rewarding those with real estate while exacerbating poverty among those without. George offered the solution of a single tax on land and its income, a controversial idea that helped make him famous but did not bring resolution to the problem.

Big landowners dominated the Central Valley for several reasons besides their ability to take advantage of loopholes in public land laws. In some ways, valley conditions actually favored size over subdivision. "Industrial" farms like those of Glenn in the Sacramento Valley and Miller & Lux in the San Joaquin developed because there were in fact some natural impediments to small operations and some factors that encouraged large ones.[10]

The physical conditions of much of early California, for example, particularly in the Central Valley, worked against small landholders. The variable soils and uneven distribution of water discouraged the formation of a patchwork of small farms like those existing in other parts of the country with more reliable rainfall. Years of experience might be necessary to work out the best adaptations to Central Valley conditions: summer drought and blistering summer heat, variably rainy winters and occasional disastrous flooding. On the dry valley plains farming had to move by rhythms different from the East or Midwest, and experiments might fail. Private irrigation systems were difficult to build and too expensive for most individuals. On the other hand, in the river bottoms and delta marshes, masses of tule grass could frustrate efforts to farm, and reclamation activities could exhaust both labor and capital. It would take decades, in some places nearly a century, before public reclamation or flood control and irrigation systems would be available for farmers.[11]

Access to capital for farmland development was thus a necessity in the nineteenth century, as it is even now. Few farmers could start on a shoestring, and none could survive without some kind of financial cushion to absorb the blows of natural catastrophes or poor harvests. Successful farmers had to have deep pockets, or associates who had them.

The isolation of early California also worked against the small farmer. The sheer distance from major markets and suppliers influenced transportation costs, and here too the big men were at an advantage. Hugh Glenn, for example, chartered his own ships to avoid the high costs of shipping controlled by middlemen. And Miller & Lux got discounts for sending trainloads of cattle to market because it was more convenient for the railroad to deal with large customers than with small.

Labor scarcity also affected farmers. American farm workers were often in short supply in rural California. Though Chinese crews helped solve the shortage, their availability was an advantage mainly to those who could employ them in gangs. Until the 1880s the Chinese accounted for up to half the labor force in agriculture, but their presence undercut the ability of small landholders to farm competitively. Large-scale projects like the conversion of tule land to farmland, for example, entailed damming, ditching, and burning great masses of marsh vegetation—operations usually carried out by large groups of contracted Chinese workers who also cultivated the new fields. Big landlords bankrolled these activities, not small farmers.

The "Terrible Seventies" were a tumultuous time. "By a cruel paradox," writes one historian, "the completion of the first transcontinental railroad not only failed to bring California the expected surge of prosperity, but marked instead the beginning of a deep and general depression that continued through the whole of the next decade." [12] After the Panic of 1873 hit the west coast, San Francisco business and real estate activity turned sharply downward. Chronic urban underemployment and the glaring disparity between the rich and the poor bred popular discontent that turned into political radicalism. Mining speculation that had fueled the rise of the Bank of California now threatened to collapse it, and in 1875 its president, William C. Ralston, was forced to close its doors after account holders surged in to deplete the bank's coffers. The day after the closure, Ralston died under mysterious circumstances in the waters of San Francisco Bay, a suspected suicide.

The precipitous rise of the Southern Pacific Railroad changed life in California during the 1870s. Though the new rail lines seemed to promise the expansion of commerce, the company's unrelenting drive toward control of all public transportation was manifested in price gouging and political corruption. Resentment against the railroad magnates grew in the press, while the state began to talk of regulating rates. Populist organizations grew more active. The Grangers, who had tried to break the control of wheat brokers over farmers, now threw the new University of California into turmoil by attacking its College of Agriculture for being elitist and calling for a separate institution to offer "practical" rather than scientific studies.[13] In San Francisco Denis Kearney and his associates founded a radical new political party called the Workingmen's Party.

Toward the end of the decade Californians vented some of their dissatisfaction by crafting a lengthy new state constitution. The Constitutional Convention of 1878 brought 152 delegates together through the fall and winter to rewrite the state's primary package of laws in an effort designed to reign in abuses of power and strengthen individual rights. A strong agrarian influence predominated. The new constitution created a State Railroad Commission to oversee rail company practices and a State Board of Equalization to oversee taxation. It placed restrictions on certain corporate and legislative powers, and even made lobbying a felony (for a time). It included such detailed regulations as a mandated eight-hour day for labor on public works.

Not surprisingly, many of the new mandates were controversial. A New Constitution Party was formed to advocate for the reforms. Finally, following a heated campaign for ratification, voters approved the new constitution in May 1879 by a margin of 77,959 to 67,134, with the vote varying widely by county but heavily affirmative in the agricultural counties. Although critics called the document an egregiously inept example of democracy, the new constitution became effective on January 1, 1880. Since then it has been amended hundreds of times.

In 1879 Hugh Glenn ran for governor. Like his influential friend Marion Biggs, who was twice elected to the State Assembly, Glenn had many political contacts and was well known for his huge wheat ranch. By this time too, his success established, Glenn had begun wearing a Prince Albert coat with a fancy waistcoat and a soft wide-brimmed hat like that of a southern gentleman. With the encouragement of his friends, Glenn's own mounting ambitions led him to accept *two* nominations for governor—that of the New Constitution Party in June 1879 and, a month later, that of the Democratic Party. The Republican Party chose George C. Perkins as its candidate, while the Workingmen's Party nominated William F. White. Despite Glenn's apparent popularity as the candidate of two parties, the gubernatorial campaign turned into a classic political brawl. Glenn did not emerge unscathed.

It wasn't easy to be a political candidate in the 1870s. That decade saw the rapid succession of four men in the governor's chair. Partisanship was relentless, new parties and coalitions formed spontaneously around volatile issues, and political campaigns were vicious. Some good candidates were sidetracked by malicious critics, while others surfaced only briefly to disappear under heavy attack. The public-spirited John Bidwell, who had been elected to Congress in 1864 for a single term, was a candidate for governor in 1875 but came in trailing third after receiving scathing abuse for some of his stands—including his very unpopular support for the temperance movement.

Glenn did relatively little campaigning during the two months of the electoral battle, which took place while much of the wheat harvest was still being brought in. He seems to have paid relatively little attention to the differences between the two parties, although he displeased the Democrats when he said that he would support the New Constitution Party's positions preferentially when the platforms differed because he had been nominated by that party first. In one speech printed in the *San Francisco Chronicle*, which endorsed him in July, Glenn stated that railroad rates should be lowered, taxes cut, and monopolies checked—safe enough statements in the climate of the time. In supporting the new constitution with its regulatory laws, Glenn did in fact risk the anger of the state's banking institutions, and several of the banks that he did business with were said to have confronted him about his support for banking reform. To his credit, though he owed them $700,000, he did not retreat from his position.

This portrait of Hugh Glenn in the 1870s shows him with a long red beard and a hat befitting a gentleman.
Courtesy of Willows Museum.

The campaign drew national notice. In July the *Chicago Tribune* published an admiring profile of Dr. Glenn as a candidate for governor, noting that he was not just a "gentleman farmer" but an active manager, with hands "hard from actual labor" and a face "brown from real exposure to the sun." The *Tribune* described the Jacinto farm as productive to its capacity, a "vast sea of nodding grain" on land honestly acquired. Glenn was said to have seven foremen serving as his "lieutenants," while his livestock numbered as many as 2,000 horses, mules, brood mares and young stock. On August 3 the *Tribune* described the California election as a "bitter political war" between railroad interests and the agricultural/mining majority, who hoped to reduce heavy rail rates by at least 25 percent. The *Tribune* predicted a Glenn victory.

On August 2, however, *The Sacramento Union* came out strongly against Hugh Glenn, publishing several critical articles from other papers referring to him as a land monopolist, a "two-horse rider," and the "Duke of Jacinto." With this volley, the gubernatorial campaign turned into warfare. Because Glenn had partially repudiated the Democratic Party when he chose not to support its entire platform, the editor of the *Bakersfield Gazette* declared in true nineteenth-century style that the newspaper would "turn from him in loathing and contempt." The *Gazette* published a long synopsis of a Sacramento lecture that described land monopoly as the state's greatest problem, citing statements from Hittell's 1872 book that more than a million acres of California land were tied up in eight proprietaries alone, that Miller & Lux controlled more than 228,000 acres and that Glenn himself had 132,000 acres including his Nevada and Oregon properties. The paper sarcastically dubbed Glenn the "Lordly Earl of Colusa," who owned a thirty-mile stretch along the Sacramento River some five miles deep (true when reckoned along the meandering river bank, but closer to sixteen miles by the meridian). The *San*

Francisco Alta wrote that Glenn had no experience qualifying him to serve as governor, that he was a "pernicious" monopolist, and that the land he farmed ought to support 300 family farms.

It was during this 1879 campaign that Denis Kearney, the head of the Workingmen's Party, was at the peak of his short-lived career as a demagogue and rabble-rouser. Kearney[14] was a gifted but vituperative speaker adept at exploiting the discontents of laboring men. Himself an immigrant Irishman, he galvanized popular sentiment against Chinese immigrants. He was able to mobilize large crowds to listen to his "sandlot" speeches in San Francisco, filled with exaggeration and innuendo about the characters of the rich and powerful. Attacking the economic and political power of capitalists, Kearney and his colleagues stirred up antagonisms between landlords and tenants, employers and workers. He accused the candidate Glenn of establishing a "slave plantation" in the heart of California. With his friend John J. Doyle, a leader of disgruntled squatters in the San Joaquin Valley, Kearney spoke to large crowds condemning the Southern Pacific and its control over vast tracts of land. Their firebrand oratory was at least partly responsible for the squatter rebellion that surfaced during the Mussel Slough incident of the following year.

According to the pungent prose of his critics, Hugh Glenn used "brutal methods to obtain his covetous ends." He was "too cunning and selfish to do his own farming" and rented three quarters of his land to tenants who had to assume all the risks while he got all the profits. In his "miserable grasping career" Glenn was running a kind of Irish tenant system in which he controlled his lessees by giving them loans at 18 percent interest with money he himself borrowed at 9 percent. His employees had to work under brutalizing conditions and were used like beasts, working 16-hour days at $1.50 day or about 9 cents an hour, and when they cashed their checks at the company store, he took back a percentage. Worst of all, they said,

Glenn preferred Chinese labor to white, thus depriving American men of work. One newspaper article referred to him as "Ah Glenn" and claimed that his sons were learning Chinese.

The *Sacramento Daily Record* extensively criticized Glenn's ranch management. This newspaper ran a long article on "How Dr. Glenn Farms His Lands" beside a letter by an alleged former employee castigating labor conditions at Jacinto. An "interview" said to have been conducted with a nearby landowner blasted Glenn for operating so selfishly that the entire county lacked social amenities and remained primarily a "barren waste' that could otherwise be developed into small or mid-size farms. And, although Glenn was not alone in growing wheat, he was accused of depleting the soil by concentrating on a single crop.

A particularly unpleasant story surfaced in the *San Francisco Post,* describing Glenn's alleged relationship with a Matthew Murray in Colusa County. According to the story–which was said to have been told first by Denis Kearney–Glenn had tried to buy Murray's 160-acre farm property in early 1874, but the Irishman had refused to sell it. Later that year Murray, a man of modest means who also ran a small saloon out of his house, became involved in a violent scuffle with a drunken customer, a quarrelsome 19-year-old who rode his horse up onto Murray's porch, shouted insultingly at his wife, brandished a weapon, and would not leave when told to do so. In self defense Murray had fatally shot the man. After the incident Murray had turned himself in to the sheriff, and after a court hearing he had been discharged. According to the *Post*, Glenn subsequently pressured the authorities for a jury trial, at which Murray was convicted of murder in the second degree and sentenced to a term at San Quentin Prison. To pay his court expenses, he had to mortgage his farm. While he was serving his term, Glenn arranged to buy his land from the mortgage agent.

When Murray came up for possible parole in 1877, the story

said, Glenn pressured the authorities to deny it, fearing that Murray might take revenge. Murray tried a second time for parole during the year of the election, enlisting the support of dozens of Colusa County residents as well as that of an eloquent Catholic priest and most of the jurors who had convicted him. Petitioning for his release, supporters cited "mitigating circumstances" regarding the shooting and the destitution of his wife and child during his imprisonment. The judge in the case and the defending attorneys also supported the petition. The prosecuting attorney was ambivalent, stating that he believed there had been a "great deal of perjury" during Murray's trial, and that although he would not sign the petition, he also would not object to Murray's release. As the plea for clemency went forward to the governor, the *Post* story appeared, blaming Glenn for Murray's imprisonment and portraying him as a mean-spirited man bent on fulfilling his ambitions as a land baron. The story was in line with the paper's general bias. No version from Glenn's perspective ever surfaced.

After a contentious two-month campaign, full of nasty attacks and innuendo, Hugh Glenn lost decisively in the September 1879 election, defeated by Republican George C. Perkins, 67,970 to 47,650. Of the total statewide count Glenn received somewhat less than a third. He won the majority in his own county and in several other rural areas, but the urban counties voted heavily for the other candidates. The Workingmen's candidate received 44,480 votes, slightly below Glenn's final tally. In retrospect, Glenn's employment of Chinese laborers was said to have been one of the chief reasons for his defeat.

As for Matthew Murray, the new governor commuted his sentence in January.

---

1 Descriptions of Glenn's farming practices are found in many contemporary sources, including newspapers, the 1880 national census of agriculture, and county histories.

2 Wik, "Some Interpretations of the Mechanization of Agriculture in the Far West," *Agricultural History* 1975.

3 "Cyrus Hall McCormick," Library Archives, Wisconsin Historical Society.

4 Davis, "Wheat in California," *Overland Monthly*, 1868.

5 The painting was displayed in public and became well known but later disappeared. It was rediscovered in the 1940s in a vacant farmhouse near Corning by agricultural historian F. Hal Higgins. The *Pacific Rural Press* reprinted the original engraving with a story by Higgins in October 1946. The painting now belongs to a Hoag descendant.

6 Street, *Beasts of the Field*, describes the life of farm laborers in detail, taking a consistently grim view.

7 See Sucheng Chan, *This Bittersweet Soil: The Chinese in California Agriculture, 1860–1910.*

8 Nordhoff, p. 139.

9 See Kelley, *Battling the Inland Sea*, and *Gold vs. Grain: The Hydraulic Mining Controversy in California's Sacramento Valley.*

10 Gregor, "The Industrial Farm as a Western Institution," *Journal of the West*, 1970.

11 In the early 1870s serious discussions took place in California about the feasibility of comprehensive flood control and irrigation systems. In 1874 the Alexander Report, commissioned by an act of Congress, proposed such a system for the Central Valley. Eigenheer, pp. 320–24.

12 Rawls and Bean, p. 165.

13 Scheuring, *Science and Service*, pp. 14–19.

14 Dennis Kearney was no relation to Theodore Kearney.

~6~
Death at Jacinto

BY 1880 HUGH GLENN WAS VERY WELL KNOWN. Though he had lost the election for California governor, he was widely recognized as the state's leading wheat producer, and as his outsized operations grew, they drew fascinated attention. Journalists touted the Jacinto Ranch as the most productive wheat farm in the world. The *Chicago Tribune* ran another story on Glenn and his huge enterprise in September 1880, and other national reports featured him as well. Most of them focused awestruck on the army of workers and machines required to bring in the crop. The *Tribune* story read like a farm inventory, reporting on the animals, the men, the machinery, and the buildings that made up the overall business, and Glenn was portrayed as a meticulous manager who visited every section of his holdings at least twice a week.

The Glenn land now totaled about 66,000 acres in the Sacramento Valley. Jacinto was a well-organized little company town with a complex of buildings ranging from the hotel, saloon, and post office to warehouses, barns and slaughterhouses. Chinese gardeners living along the river tended some fifty acres of fruits and vegetables. Glenn's wheat acreage consisted of about 55,000 acres; he told a journalist that he leased out about 14,000 acres on shares, farming the rest himself. His two largest fields stretched for seven and a half miles and took three days to complete a round with team and plow. To feed the thousands of animals on his holdings he also grew barley, oats, and wild hay. To plant and harvest his huge crops he owned dozens of gang plows, harrows, headers, and wagons, and he used sixteen threshing machines and a thousand mules at the peak of harvest season. Fourteen blacksmith shops employing sixteen blacksmiths were said to be kept busy building and repairing the equipment. His responsibility for managing all the logistics was shared partly with his tenants, who were to receive one half of the crop income after expenses.

The Jacinto Ranch also received nation-wide publicity in New York journalist *Frank Leslie's Illustrated Weekly Newspaper.* This popular publication reported with awe that Glenn was compelled to hire as many as 50 goose hunters to protect his crop from wildlife depredation during the winter growing period, men on horseback using shotguns to kill incredible numbers of birds.

The 1880 federal census report on national cereal production included a section on Jacinto too. Although California ranked only seventh that year in total wheat yield, behind six Midwestern states, California farming practices rated a special section in the report. The census identified Colusa County as the state's most productive county, accounting for 4.5 million out of an estimated 29 million total bushels of production. Hugh Glenn was named the largest producer in the state.

The census report also described Glenn's farming operations in minute detail. After fallowing, eight-mule teams would pull gang plows into the fields to prepare them for sowing with bluestone treated seed, and sowing would take place throughout the fall. During the growing season the fields would be patrolled by

FRANK LESLIE'S ILLUSTRATED NEWSPAPER

Entered according to Act of Congress, in the year 1882, by Mrs. Frank Leslie, in the Office of the Librarian of Congress at Washington.—Entered at the Post Office, New York, N.Y., as Second-class Matter.

No. 1,410.—Vol. LV.] NEW YORK—FOR THE WEEK ENDING SEPTEMBER 30, 1882. [Price 10 Cents.

CALIFORNIA.—THE EXTERMINATION OF WILD GEESE IN COLUSA COUNTY—A PATROL-GUARD FIRING INTO A FLOCK OF THE FOWL.

Frank Leslie's Illustrated, a weekly newspaper in New York, ran a story describing the wholesale slaughter of wild ducks and geese in the Sacramento Valley.

Courtesy of California Historical Society, North Baker Research Library, San Francisco.

goose hunters while equipment was readied for harvest. When the grain was ripe, headers would be sent out to cut about 100 acres in the center of each field, the cut area then serving as the site for a large stationary thresher/separator driven by a 44-inch cylinder, 26-horsepower engine. Six-mule teams pulling 16-foot headers could average a harvest of about 30 acres a day. Immediately after being headed and delivered by wagon to the separator, the grain would be threshed and separated. Glenn's two gangs of 35 men and 84 animals at the headers and 32 men and 11 animals at the separator added up to a total of 67 men and 95 animals for a good day's harvest work crew. When things were running well, the average time for the grain to move from field to sack was said to be just 15 minutes. The total wheat harvested a day on the Glenn ranch was said to be an average of 3,800 bushels or 1,700 sacks, which would be hauled to the riverbank to be piled onto barges.

According to the census, the average yield for the entire Glenn ranch in 1879 was 19 bushels per acre–not a particularly good yield, though the 1880 crop was expected to be better. Generally, in more favorable years, Glenn could count on an approximate 25-bushel yield, a good average for mid-valley soils. On better land, wheat yields could go much higher, even to 60 or 70 bushels to the acre. John Bidwell once reported a record yield of 73 bushels per acre on 10 acres at his Chico ranch, and cases were on record of yields above 80 bushels on freshly opened land.

Closer to home, a *Sacramento Union* reporter writing about the more human side of Glenn's operations described Jacinto as a "pleasant little village" with an "exceedingly home like appearance," where Glenn was said to stroll about daily and lend an occasional hand at various activities. This reporter depicted Glenn as a "quiet and unassuming" individual who managed his farm himself and worked very hard for his success, a man with "sun browned face and horny hands," loved and respected by all—a portrayal somewhat at odds with that given to the public by Glenn's opponents in the 1879 election campaign.

Wheat prices dropped to $1 a bushel in 1880, and Glenn looked for ways to cut his costs. To avoid heavy brokers' commissions he himself chartered twelve ships to carry his crop. Selling the grain by telegraph, he exported 27,000 tons to England that year. Two freight trains of thirty to forty cars each left Willows daily that fall loaded with grain, most of it from Jacinto. For that year's harvest Glenn received a reported $800,000, much of it destined to pay off accumulated debts. His new ambition, he said, was the production of one million sacks of wheat in a single year—a goal he never quite reached. Glenn's tenant George Hoag, meanwhile, who farmed some 13,000 acres for Glenn, filed for bankruptcy the following year, a victim of crop failure and the 18 percent interest charged for cash advances.

Glenn began to look into upgrading his field equipment to save on labor costs, testing one of the big new steam-operated combine harvesters from Stockton. Combines were designed to further cut the need for workers by joining all cutting, heading, and threshing functions into a single machine that would eliminate the need for headers, wagons, and stationary threshers, dropping filled sacks onto the ground at the end of the combine process. Pleased with the test, Glenn sent one of his stationary threshers down river the following year for conversion to a combine for use in 1882. Though the cumbersome machines were costly, Glenn calculated significant savings on labor as well as greater reliability in the long run. Machines in the shed ready to roll would at least partially eliminate the problems of labor recruitment, always a concern as harvest neared in this relatively remote region. Other growers who thought the same way and could afford it bought combines as soon as they became available.

Despite the advantages of the new machinery, many growers stuck with mule-drawn harvesters, for steam engines always posed a fire hazard in the dry fields. Mules were still generally preferred as draft animals because they were tougher than horses, had hardier feet and more stamina, and could handle the heat better. Horses had to be more frequently rested and cooled, requiring water to be hauled out to the fields in tanks up to five times a day. The use of mule teams in some places continued throughout the wheat era, especially in the rolling hills where top-heavy steam combines might tilt dangerously.

Glenn and others who took advantage of new harvesting machines must have been pleased with their greater efficiency. From a pragmatic point of view, machines saved time and money, raising profits. From another point of view, however, machines were alien intruders in a pastoral landscape. Literary men especially found it hard to relate to the large-scale mechanization of farming. Young novelist Frank Norris expressed something like dread in his typically extravagant description: "The harvester, shooting a column of thick smoke straight upward, vibrating to the top of the stack, hissed, clanked, and lurched forward. Instantly, motion sprang to life in all its component parts; the header knives, cutting a thirty-six foot swath, gnashed like teeth; beltings slid and moved like smooth flowing streams; the separator whirred, the agitator jarred and crashed; cylinders, augers, fans, seeders and elevators, drapers and chaff-carrier clattered, rumbled, buzzed, and clanged. The steam hissed and rasped; the ground reverberated a hollow note, and the thousands upon thousands of wheat stalks sliced and slashed in the clashing shears of the header, rattled like dry rushes in a hurricane, as they fell inward, and were caught up by an endless belt, to disappear into the bowels of the vast brute that devoured them."[1] (Unlike Norris, farmers and mechanics were more likely to view these mechanical "brutes" with fascination at the sheer cleverness with which they were put together.)

In 1880–81, John Hittell reported 356 ships taking wheat, flour, and barley out of the Golden Gate—191 of them British, 123 American, and the remaining other European—well over half going to Liverpool, others bound for the Sandwich Islands (Hawaii), China, Australia, and South America. Many of these ships were loaded in San Francisco (103), but 97 left directly from Vallejo, 84 from Port Costa, and 31 from Benicia. Freight rates that year to Europe averaged $15.75 a ton, no small matter for a cargo that was a wholesale commodity and in part a speculative venture. The following year, 1881–82, hit a record of 550 ships sailing out of the Golden Gate carrying wheat and flour, nine more with flour alone. That year the ships were registered as 405 foreign, mostly British, and 154 American.

Some ships now had a new beginning point in their lengthy journey across the seas, a loading facility and community named Port Costa along the south side of Carquinez Strait, where the rivers of the Central Valley mingle to flow into San Francisco Bay. The strait is a mile and half wide where the rivers meet tidewater, and sheltered to some extent from coastal weather by the enclosing hills. In September 1878 the Central Pacific Railroad bought a right-of-way along the south side of the strait and built spur lines into the area as part of its new construction between Stockton and Oakland. The company also bought out the California Pacific line running from Sacramento to Vallejo and rerouted that line to run from Sacramento through Benicia to Oakland via a ferry transfer at Port Costa. Port Costa was founded in 1879 as the landing for the new railroad ferry, which would now be a link in the transcontinental railroad's more direct route to its terminus in Oakland.

The railroad ferry, named the *Solano*, was designed to transport entire trains across the strait, carrying only rail cars and foot passengers. It was built to ambitious specifications in West Oakland. At the time the world's largest such ferry, its wooden hull was 425 feet long, with a 116-foot beam, and it parted the waters with powerful 2500-horsepower twin walking beam engines. It had four railroad tracks on deck and could carry up to 36 rail freight cars or 24 passenger cars, plus two locomotives for switching.

In November 1879 the ferry was ready for service. On its first official run, December 28, 1879, from Port Costa to Benicia (then a military base), the ship was greeted with a 21-gun salute. This ferry would operate for 51 years with very few problems, taking less than an hour from loading of the rail cars to reassembly and departure, and shaving more than an hour off the previous roundabout rail route from Sacramento via Stockton into Oakland. As the Central Pacific's main line into the Bay Area, it would operate twenty-four hours a day.[2]

Railroad magnates and grain merchants soon realized that there were many advantages to loading ocean-bound cargo at the north end of the Bay rather than in San Francisco itself. Port Costa became established as a deep-water port along Carquinez Strait to carry on the transfer of wheat to merchant sailing ships bound overseas. Businessman George Washington McNear bought much of the long frontage between Crockett to the west and Martinez to the east to lay out new docks specifically for this purpose. Shore tidewater depths along this strip ranged from 25–50 feet, and the sheltered location around the bend from the Bay and along the narrow curve beneath tall hills could now combine both river and railroad traffic with ocean shipping.

From Port Costa it was 30 miles yet to the ocean, but it was less expensive and more efficient to load grain ships there than in the city of San Francisco. The spot became a major grain loading

The big *Solano* railroad ferry transported entire trains across the Carquinez Strait for decades before a railroad bridge spanned the gap.
Courtesy of San Francisco Maritime NHP Library.

port in the 1880s, surpassing all other ports. McNear opened his Port Costa Warehouse and Dock Company in 1881, and the British firm Balfour and Guthrie built its California warehouse there the same year. Soon several other company warehouses and long wharves were joined by seven hotels, a dozen saloons, and a conveniently situated brewery to slake the thirst of workers. At the peak of the shipping season the little town of Port Costa might hold as many as 3,000 inhabitants, mainly sailors, stevedores, and railroad employees. The stevedores who handled the cargo—hundreds of thousands of 100-pound sacks—were big, burly, rough characters, many of them immigrant Irish who in other seasons might work in the valleys on harvest crews or in the cities on construction. Port Costa acquired the nickname "Corned Beef Junction" for the stevedores' favorite food (or sometimes "Port Chaw" for their use of tobacco), and it gained a reputation as a colorful, sometimes violent community.

At its peak Port Costa took in more than 50 trainloads of grain a day as well as varying amounts of barge cargo. The grain companies could load up to a dozen vessels at a time out of the long, low warehouses where the grain was cleaned and graded. Only two years after opening, Port Costa sent out more than a hundred ships a year, several a day during the shipping season, and the number increased annually for the next decade. By 1887 four fifths of the Central Valley's wheat trade was going out through the Port Costa waterfront, and for more than 20 years the long wharves sent millions of dollars in exports out through the Golden Gate. Even in 1900, as the trade was declining, Port Costa loaded some 30 million sacks.[3]

The Glenn family acquired a new home in Oakland in 1881. Despite his considerable debt, taken on through his many expansions and improvements to farming operations, Hugh Glenn was doing well

Port Costa grain warehouses shipped countless tons of California wheat to distant international markets for well over twenty years.

Courtesy of San Francisco Maritime NHP Library.

enough financially to buy the former James H. Latham residence at 1350 Jackson, a Victorian mansion near the corner of Jackson and Lake streets in an affluent neighborhood overlooking Lake Merritt. This grand home had three stories, two turrets, a music room, a ballroom, a conservatory, a billiard room, and a bowling alley. Fronted by the popular palm trees of the day, it also had a generous garden in back, complete with a fountain and stables. The home had cost $125,000 to build in 1873, a large amount at the time, but after the death of its owner Glenn was able to purchase it for only $65,000. The house provided a prestigious address for his wife Nancy, who now began to spend much of her time away from Jacinto overseeing their family of growing children, several of whom attended schools in the Bay Area. Ella Glenn, their only surviving daughter, was enrolled in Mills Seminary for Women, a well-regarded boarding school not far from the new home.

No one knew it then, but 1882 marked the high point of wheat production in California, with 1,128,031 tons exported abroad. Prices were also very favorable, for terrible weather in England and Europe in 1881–82, combined with a severe winter on the Great Plains in North America, kept worldwide demand high. Though it was a banner year for California wheat in general, however, Glenn's own production was down, and he had other problems. That year saw a dry spring in the Sacramento Valley, and as the crop began to mature, a fire destroyed most of one of Glenn's large wheat fields, with the standing grain reported to be worth $100,000. Although Glenn apparently took this calamity calmly, his 1882 harvest amounted to only half the anticipated yield. To add to his financial pressure, Glenn's workers went on strike during harvest season for a 30-cent-a day raise. Glenn had to mortgage some of his land to pay the bills.

There were compensations, however. A major family event was the winter marriage of Glenn's 22-year-old daughter Ella, who had recently graduated from the Mills Seminary in Oakland. On February 1, 1883, in a festive wedding at the Glenn home in Oakland, she married Peter French, the manager of Glenn's huge cattle ranch in southeastern Oregon. The celebration was an elegant affair, reported in the society pages of both the *San Francisco Call* and the *Oakland Tribune*. The parlors were beautifully decorated, the guests were mainly family and close friends, and the bride was "attired in a crushed strawberry dress of corded silk and brocaded satin, adorned with point lace and tasteful flowers." The bridal couple undoubtedly made a handsome pair, despite their differences.

Peter French was now 33 years old, a small, wiry, dark-haired man who had earned a sterling reputation as a cattleman. He was quick, energetic, and full of drive; he knew the cattle business and the Oregon back country; and he was a born leader of men. Ella Glenn was better educated and accustomed to gracious living, but she had never been out of California, certainly never in the remote high-desert region across the northeast border. Peter had come to know Ella on his annual visits to California, but the pair's brief courtship had hardly allowed them to learn much about each other. After the wedding the two set out for a honeymoon trip through California, mainly visiting relatives and friends at Jacinto and Red Bluff, followed by their departure for southeastern Oregon through the Winnemucca route. But in Winnemucca they were abruptly recalled to California by some terrible news.

Murder suddenly ended Hugh Glenn's career on February 17, 1883. He was 58 years old and had spent years climbing toward fame and financial success. Now a shotgun blast took his life and tarnished his reputation. The month had begun with family festivities at Ella's wedding and a respite from the demands of ranching, but after the celebration Glenn had returned to Jacinto. A scant two weeks later he was dead. The outlines of the story were clear, the details less so.

At 2 in the afternoon Glenn was standing on the Jacinto hotel porch, watching his men catch a troublesome horse while he waited to walk to the stable. A former employee, Huram Miller, who had recently been hired and then dismissed as the Jacinto bookkeeper, approached the hotel porch carrying a loaded shotgun. Apparently without warning, from 12 to 15 feet away, Miller lifted the gun and fired at Glenn, hitting him in the head above and behind his right ear, tearing away some of the skull. Mortally wounded, Glenn fell onto the porch. Stunned by the noise of the blast, people poured out of the hotel. Women screamed while men came running from the yard to try to help. Glenn was carried into the hotel while someone went for a doctor, but it was clear that he was dying.

Meanwhile, the gunman ran from the scene up the river road and then cut through a nearby field heading toward Willows. The farm superintendent and several others took off after him on horseback and by buggy, firing warning shots and shouting for him to drop his gun. He did not do so until one pursuer finally shot him in the leg just above the knee. The injured man fell to the ground, only then throwing away his weapon, and was captured. Before a mob of excited bystanders could lay hands on him, his pursuers took him to Willows to be placed under arrest while his wound was dressed. He was then taken to Colusa, the county seat, to be delivered to the authorities. Back at the Jacinto hotel Glenn lived only a few more hours, never regaining consciousness. Attended by those family members near enough to come to his bedside, he died that night.

Three days after Glenn's death his body was transferred from Jacinto to the railway station at Willows. In the largest memorial procession in local history, hundreds of horse-drawn vehicles solemnly moved along the country roads following the hearse bearing the coffin. The formal funeral ceremony took place in Oakland, attended by dozens of extended family members and many prominent citizens including Governor Perkins, who served as a pall-bearer, two other former governors, several judges, Charles Lux of Miller & Lux, and a host of other political and social leaders. After the funeral his family interred Glenn's body temporarily at Oakland's Mountain View Cemetery, later moving it to the family cemetery at Jacinto.

In the days and weeks after Glenn's murder, the *San Francisco Call* and other newspapers covered the story. The *Call* reported in grisly detail that the gunshot had taken out a piece of the victim's skull along with a part of his brain as big as an egg. The *Chicago Tribune* described the event as the "brutal murder" of a prominent Californian who had started his career as "originally a horse-dealer." One San Francisco paper printed a scurrilous attack on Glenn after his death, but the *Pacific Rural Press* called him a "much esteemed" member of his community. Some eulogies were fulsome, as in the *Oakland Tribune*: "Dr. Hugh J. Glenn, of Colusa, was one of the remarkable men of his age. He was born to be a leader in whatever he undertook. Had he commenced his business career in Wall Street, he could have been another Commodore Vanderbilt or Jay Gould....Educated as a physician, an unexpected event made him a land holder, and in a few years he became the largest farmer on the globe, and the only wheat grower in America who chartered his own ships and sold by telegraph his own grain to Europe, dispensing with middlemen entirely."

On March 6 Huram Miller appeared in Colusa County Court to be charged with first-degree murder. The trial was set for June. The murderer, age 48, was a small, blue-eyed man struggling with serious alcohol and money problems. Born in northeastern Missouri not far from St. Louis, Huram Miller, like Glenn, was a transplant to California. Though there was 10 years difference in their ages, Miller and Glenn had been acquainted in their youth, and their wives were friends; Glenn had been kind to him because of that. Miller had immigrated to California in 1874 and lived for a time in the town of Colusa, but despite his good education and business experience he had been unsuccessful in attempting to start a business there. In 1881, with Glenn's backing, he opened a grocery store, but this venture also failed after a year and a half, at least partly because of Miller's weakness for alcohol. While his wife taught school to help support their five children, Miller worked briefly for a rancher and herded geese near Willows, meanwhile suffering from deep depression. In January 1883 Glenn offered him a job as bookkeeper and cashier at Jacinto on a trial basis. Glenn had asked outright if Miller could stay sober, and Miller had told him that he had given up drinking a few months before.

It had been a hard year for Huram Miller, and January was an especially hard month. In September 1882 he had been forced into county court for insolvency, where a hearing took place to determine whether he could be declared bankrupt in order to stop being pursued by his creditors. Miller was ordered to prepare an inventory of all his possessions, and the judge had directed the sheriff to take over all assets so that Miller's creditors could be paid some percentage of what he owed them. Notice was published in the *Colusa Sun* that the court would hold hearings for the creditors, and in January 1883 Miller was declared bankrupt and ordered to pay a small percentage on his creditors' claims. He then applied to be discharged from all his debts, and the court ordered that creditors had to show cause why not the following month. Thus Miller's bankruptcy coincided with Glenn's offer of a new job; but Miller soon found that the new job itself was more stressful than he could have imagined.

According to Miller's later testimony, Mrs. Glenn was living in

Oakland at the time he started his job at Jacinto in January. The old house had been converted into lodging for hired men, and the "new" Glenn house into a boarding house for others, including office employees like Miller himself. Miller began his work at the business office in the Jacinto hotel building. In his job as bookkeeper he had to work closely with a Mrs. C. B. (Carrie) Posten, also known as "Rosebud," who managed both the hotel and the boarding house, where she was also living. Mrs. Posten was a widow, said to be a beautiful woman of French extraction with a fondness for luxury, and rumored to be close to Hugh Glenn. Miller described her as an influential and arrogant woman who was handsomely paid for her managerial duties, earning $150 a month as well as the profits from the hotel and saloon.

Mrs. Posten was also authorized to cash ranch employee checks, taking a percentage for this service. Most ranch employees had little opportunity to cash their weekly or monthly checks at banks in town and depended on getting cash in Jacinto. When the men presented their checks for cash, the bookkeeper first subtracted their expenses as noted on the store and saloon books, and then Mrs. Posten took a five percent discount for managing the funds. Miller soon discovered that the men resented this discount, especially because they knew that she earned far more than they did (the average working man's monthly wage was then around $30). According to Miller, when he told Mrs. Posten about the complaints she grew irritated and said she ought to be getting even more for handling the cash responsibilities. Disturbed by the employee grumbling, Miller was also disturbed by Mrs. Posten's proprietary attitude toward the business office, as well as by her disparaging remarks about Mrs. Glenn. Feeling stress, Miller began drinking again.

In his deposition before the June trial, Huram Miller stated that he was broke and that his brother in Texas was paying his expenses and sending a lawyer to assist Jackson Hatch, the defense attorney. Hatch asked the court for a postponement of the trial to July, but did not get it.[4]

The June trial proved unsatisfactory. Of seventy-five prospective jurors called, dozens were challenged for their "state of mind" and bias about the case. It took days to assemble a group acceptable to both sides. After the jurors were finally seated, numerous witnesses gave conflicting testimonies, and some of these were struck out of the record. Some witnesses said that the shooting was sudden and unprovoked, while others testified that Miller had been involved in a loud argument with Glenn a week before and that he had been carrying a shotgun for days afterward. (Mrs. Posten was not among the witnesses, for she had left the state by this time and was unavailable to be subpoenaed. She never gave her side of the story in public.)

Coached by his attorney, Miller pleaded self-defense. He testified that he had been beaten badly during an argument with Glenn, that his fear of Glenn afterwards had impelled him to shoot suddenly when he saw him later, and that the incident had been a tragic accident. He said that he had been carrying the shotgun on the day of the murder because he was trying to raffle it off to get money for his expenses, and that the gun happened to be loaded because he had been shooting at birds. He offered a raffle paper as evidence, showing seven signatures on a list of 40 chances.

Days of testimony went into jury deliberations that ended in a hung jury, 6-6, and Miller remained in jail. The judge set a second trial for September. Miller's attorney offered many procedural objections–including a plea for continuance because the defense could not bring in enough new witnesses before September, and an argument that the principle of "once in jeopardy" made a second trial illegal–but the case was reopened in September 1883.

The new trial was the talk of the county, and few of the locals were without an opinion. For this trial, which lasted fourteen days, 400 residents were screened for the twelve slots on the jury. The court files contain expense claims for at least 19 subpoenaed witnesses for the defense.

On the prosecutor's side, laborer J. D. Thornton testified that he had been standing near Glenn on the hotel porch when he saw Miller suddenly shoot Glenn in the head from about 15 feet away with a double barreled shotgun. According to Thornton, the victim had his head turned away and was watching some men trying to catch a horse. Glenn did not see the gun aimed at him, had his hands in his pockets, and no words were exchanged. Several witnesses corroborated this testimony. Glenn's son, Charles Glenn, testified that on the night of February 9 Miller had been so drunk at dinner in the boardinghouse that he had spilled soup on the table; that he had left the table abruptly; and that Glenn had followed him out of the dining room. Others testified that they had heard a loud argument afterwards in a room at the back of the house. Some witnesses recalled that they had seen Miller walking around in the following days with a bruised face and a hand in a sling, and that he was carrying a shotgun. He was said to be loitering outside the hotel with the gun during a dance held the night before the murder, looking in the window.

On the defense side, Miller's own testimony was detailed, dramatic, and carefully modulated to elicit sympathy. Again pleading self-defense, Miller placed Mrs. Posten squarely in the middle of the events leading up to Glenn's death. According to Miller, during the few weeks of his employment as bookkeeper Mrs. Posten had been "meddlesome" in her examination of his accounts and had gratuitously remarked to him that Mrs. Glenn had had an affair with the former bookkeeper. Miller said that the remark had made him angry, and he had told her not to insult a good woman. Mrs. Posten had retorted that the Glenns were going to be divorced and that she would then marry Hugh, and that Miller couldn't criticize her and get away with it.

On February 8, Miller continued, the friction between them developed into a major quarrel about Mrs. Posten's discounts in cashing checks and her opening of Glenn's mail. Miller said that he believed that she told Glenn about the quarrel and said she wanted the old bookkeeper back. The next evening Miller showed up for dinner at the boardinghouse (at which he said the servant spilled the soup) but soon left the dining room, followed by Glenn. In describing the argument in Mrs. Posten's room at the back of the house, Miller said that Glenn had angrily dismissed him from his bookkeeping job and ordered him to turn over the keys to the store and safe. When Miller told Glenn that Mrs. Posten was slandering his wife, Glenn hit him with the big brass key in his fist and broke his nose, cursing him, knocking him to the floor, kicking him in the "privates," and giving him a terrible beating while Mrs. Posten stood by with a pistol in her hand.

Bruised and stunned, Miller staggered out of the house and nursed his wounds for a few days. He sent for a shotgun that he had taken to a Chico repair shop, which he meant to raffle off for some much-needed cash. He used the gun a few times to shoot at geese, but on the night of the dance at the hotel, where he had thought he might sell raffle tickets to the guests, he had "terrible diarrhea" and left the building, remaining outside for a while. The next afternoon he again came to the hotel with his gun to sell raffle tickets, and Glenn happened to be standing on the porch. On seeing Miller, Glenn verbally abused him and grabbed for his gun—"we had a little pull"—and then Glenn put his hand in his pocket saying he would "fix" Miller. At that point, said Miller, his fear was so great that he reflexively pulled the trigger of his gun, not even aiming it, and unfortunately hit Glenn in the head. In horror, Miller had run from the scene.

After lengthy summations and interpretations of the conflicting testimonies by both the prosecuting and defense attorneys, the trial proceeded to jury deliberations. Judge E.A. Bridgford spoke at length to the jury, explaining in detail the distinction between the findings necessary for a conviction of murder in the first or second degree (an act performed in deliberate cold blood or with general malice aforethought) or for manslaughter (an act of momentary passion or under mitigating circumstances), and he reminded the jurors to consider the concept of "reasonable doubt" in passing their verdict. He also granted the jury the right to determine the penalty for the crime, imprisonment or execution.

On October 5 the jury found Huram Miller guilty of murder in the first degree, recommending life imprisonment. On October 21 the judge issued Miller's sentence, incarceration for life, and he was sent to Folsom State Prison. Although Miller's attorneys appealed for a third trial and even made a request to the State Supreme Court for a hearing, these pleas were denied.

Hugh Glenn left no will. His assets at the time of his death were estimated at well over $1,200,000, an immense sum at the time, but newspapers soon reported that his liabilities amounted to even more. High expenses and a series of misfortunes on the ranch had accumulated so much debt that the estate had barely enough income to cover even the current balances due. Although the value of his 1882–83 wheat crop came to an estimated $700,000, many creditors stood in line to receive payments. Ironically, almost immediately after Glenn's death Carrie Posten herself had filed a claim against his estate for $6,437.65, but within a month she withdrew the application and vanished from the state. Other creditors, much larger ones, were more stubborn.

Executors of the estate moved quickly to liquidate some assets. Within a few weeks after Glenn's death there was brisk bidding at the sale of several tracts of his property around Princeton amounting to several thousand acres. Before year's end more than 11,000 acres of the Jacinto ranch were sold to raise money for settlement of the estate—nearly twenty percent of all the Glenn holdings in the Sacramento Valley. Within two years the Miller & Lux partnership bought Glenn's Nevada ranch at Soldier Meadows for $250,000, and the family put the Oakland mansion up for auction.

In a string of unfortunate accidents during the following years, the Glenn family's troubles continued. S. E. Wilson, Nancy Glenn's brother-in-law and Glenn's former partner in the livestock trade, became the first administrator of the estate, but he died two years later in a road accident. (N. D. Rideout, a local banker, then became the administrator.) Glenn's survivors included his wife, five sons, and one daughter, but the oldest son, Alphonso, died in 1888 at the age of 38, apparently of a heart problem. A further blow to the family came later the same year when Hugh Glenn, Jr., known as "Major," died at age 20 in an unexplained fall from a second-story hotel balcony at Willows. Two years later, Roy, the youngest son, also died at age 20 when he fell out of a boat and drowned in floodwaters near Jacinto.

Ella, who had married Peter French only two weeks before her father's death, returned to be with her mother through the terrible weeks afterwards. Then, because she was pregnant, she remained in California to bear a son named Harold Glenn French on September 22, 1883, just as Huram Miller's second trial was underway. Perhaps because of her mother, perhaps because her short experience in Nevada had convinced her that the remote backcountry did not appeal to her, Ella appears never to have left California again to live with her husband, nor even to visit him. Peter French came to California on ranch business periodically and remained remarkably loyal to the Glenns throughout his life, yet their marriage seems to have been one of convenience and considerable distance. With no great love between them, certainly their differences were profound. Ella was born to a life of luxury and loved the city, while Peter was dedicated to the cattle business and the P Ranch. Young Harold, raised by his mother, seems to have had little connection with the man whose name he bore, and some have speculated that he was not Peter French's son at all.

In February 1891 Ella sued for divorce from Peter French on the grounds of desertion. With no objections from Peter, the divorce was granted. The court awarded custody of the boy to his mother, Ella took back her maiden name, and two months later she married her second husband, Charles Lee Leonard, a bookkeeper. This marriage produced no children.

Perhaps Hugh Glenn's reputation was tarnished by the manner of his death, but it was not totally eclipsed. His reputation as a "big-hearted" man with few enemies survived the scandal of his death enough for him to be celebrated in county histories, and when local pressure grew in favor of dividing big Colusa County into two smaller counties, Glenn's family and friends led the drive to name the new northern portion Glenn County. The proposal to divide the county had surfaced as early as 1880 in a local newspaper, which argued that residents in the north were greatly inconvenienced by the distance and time it took to do business with county offices in the south, but conditions were not then favorable for division, and the proposal languished. In 1887, four years after Glenn's death, a Colusa assemblyman introduced a bill to carve out a new county named after him, a measure that was narrowly defeated. Two years later the Legislature passed a second such bill, but the governor vetoed it because he was opposed to the proliferation of counties. It took a third try in the Legislature as well as a popular election and several court cases to complete the task. In 1891, after voters approved the measure and around the time of Nancy Glenn's death, the northern portion of former Colusa County became Glenn County, and Willows was named the county seat. But controversy over the division continued. In 1892 the court at Marysville voided charges of election fraud, and in 1893 the State Supreme Court finally ruled that the Glenn County Act was constitutional.

A second file on Huram Miller's case in the California State Archives preserves his petition for clemency in 1890. Miller's statement, painstakingly crafted in excellent penmanship, generally reiterates the testimony given at his 1883 trials: he was the victim of a vicious beating by Hugh Glenn, which left him dazed and afraid of a powerful man under the pernicious influence of a greedy woman. The inference is clear that Glenn and Mrs. Posten had an illicit relationship, and that in objecting to her practices in the business office, Miller became Posten's target. Manipulated by a vindictive mistress, Glenn punished and demeaned him, but Miller's fatal shotgun blast was an accidental, not intentional, impulse of frightened self-defense. He should not have been convicted of murder and should be released from prison.

This carefully composed plea for clemency stands as the main historical document in the case today. In the absence of countervailing testimony, it muddies Hugh Glenn's reputation. The file also shows evidence of a major effort by Miller's family to garner support for his pardon in a well-organized campaign. A petition listing 140 signatures asks for a reduction in sentence because the crime had not been premeditated; the petition must have taken some time to circulate. The signers include landowners, newspapermen, and other prominent local figures as well as five jurors from the first trial and seven from the second. Several dozen letters of support accompany the petition, many of them obviously solicited by Miller or by his daughter Katie. Some writers clearly did not know the details of the case but were responding to requests for help in circulating the petition. Many letters came from Missouri, some recalling Miller's youth and upbringing, some mentioning his wife, some stressing that Miller was "not of the criminal class." Some were from attorneys, bankers, and doctors—the "best people" of their communities urging clemency for a member of an "estimable family." Other letters simply asked for the release of a "broken down old man" who had already suffered enough.

The prison warden gave Miller a good character reference. The former publisher of the *Willows Tribune* stated that though feelings had run very high about the trial, everyone had respect for Miller's family, especially for his wife. Even W. S. Green, the well-known publisher of the *Colusa Sun* and a friend of the Glenn family, said he would not object to Miller's release. Several attorneys argued for the charge to be reduced to manslaughter since Miller had been sufficiently punished after seven years in prison. Even Glenn's old friend Marion Biggs, now a Congressman, wrote a short statement in support of Miller's release. Some letters impugned Glenn as an arrogant and overbearing character, full of "grit and push," "impetuous and spirited," bent on acquiring his fortune without consideration for underlings. Glenn was said to be dictatorial as he grew older, ruling his domain with an iron hand. One letter in the file, from Colusa Superior Court Judge E.A. Bridgford, was apparently written to Mrs. Glenn in an effort to reassure her during the campaign for pardon. Bridgford wrote that he believed the murder to have been premeditated and that clemency should not be granted. He also stated that he believed no woman was involved. Mrs. Glenn and her children forwarded this letter to the governor along with their own requesting that Miller serve out his sentence.

In the end, the Miller family efforts were successful. Governor Waterman commuted Huram Miller's sentence to 15 years on December 8, 1890, and in January 1891, having been cited as a model prisoner, Miller was discharged from prison. He and his family immediately left California, according to one source moving to Mexico.

Later that year Nancy Glenn died, survived by only two sons and one daughter out of the six sons and three daughters born to her. For the last part of her life she had lived in San Francisco with her daughter Ella, not returning to Jacinto until she was interred alongside her husband in the small family cemetery at the ranch along the Sacramento River.

The great wheat era in California in some ways paralleled Hugh Glenn's remarkable career, his murder occurring at the peak of the period. Although the immense nineteenth-century international grain trade would continue for some decades, the California share of it began to drop. The figures tell it all. In 1865 San Francisco exported a mere 1,268 tons of wheat; twenty years later that figure soared to 567,016 tons. The generally good prices that held from the 1850s through the early 1880s, however, began to decline. Wheat prices that had stayed relatively high throughout the 1870s—for most years above $1.50 per cental—started to fall steadily by late 1883. In October 1884 the price for California wheat fell below $1.00 under competition from a good English crop, and the market never fully recovered its former profitability.[5]

These declines came partly because the wars that had disrupted European harvests ended, but they were also influenced by the steady growth in international competition. During the 1880s wheat production expanded significantly not only in the Pacific Northwest but in Australia, Argentina, Russia, Canada, and other countries, which now became new suppliers to Britain and were able to undercut California prices. Sharp cuts in Liverpool prices were reflected in the returns to California farmers. Their wheat yields also began dropping in the 1880s, however, as overcropping depleted the fertility of the soils. While peak California wheat acreage stood at three million acres in 1888, peak overall production had come in 1884 with 27 million centals or 42.6 billion bushels. Historian Hugh Bancroft reported that by 1890 the average California yield was only 16 bushels an acre under the common careless methods of farming.

The bonanza years were over.

---

1 Frank Norris, *The Octopus.*

2 Matthews, "Ferryboat *Solano*," *Ships in Scale*, Jan./Feb. 2008.

3 *Sunset Magazine*, then a railroad publication, published an article on Port Costa in January 1900.

4 Records in the Colusa County Courthouse and the California State Archives are the primary sources for description of the Hugh Glenn murder trial and its aftermath. See also James, "Dr. Hugh J. Glenn: The Forgotten Legacy"; Street, "The Murder of Hugh Glenn"; and French, *Cattle Country of Peter French.*

5 Davis, "Wheat in California: A Retrospect and Prospect," *Overland Monthly*, 1898.

~7~

The Cattle King

ALMOST ALL THE LAND GRANT COURT BATTLES were over by 1870, although a few dragged on. New owners took possession of the former grant properties, many of them choice areas in the coastal region or select portions of the inland valleys, but some of the land elsewhere in the Central Valley was reserved from settlement until official state mapping was completed and the railroad companies located the routes along which they would receive grants from the federal government. By 1872 the railroad routes were determined, and more acres were then opened for claims. With its checkerboard land grants and vested interest in development along its routes, the Central Pacific Railroad (Southern Pacific after 1885) actively encouraged immigration and local improvements, heavily promoting opportunities in publications aimed at attracting settlers.

One such promotional book was Charles Nordhoff's *California for Travellers and Settlers* (1873), a publication aimed at attracting visitors. In it Nordhoff glowingly described the prospects for farming in the San Joaquin Valley, his main geographic focus, even though he also warned that the region was subject to periodic droughts and that crops might fail without irrigation. Calling Southern California, including the San Joaquin, the "best region in the whole United States for farmers," Nordhoff wrote that some of the best lands in the valley were available at just $2.50 an acre in 40- to 60-acre blocks and noted approvingly that the railroad land office in Sacramento was well organized and more efficient than the government land offices. His book clearly encouraged immigration, although its optimistic outlook, despite some cautionary notes, minimized the challenges to settlement. Newcomers with only modest resources would find that the early years in the San Joaquin were not easy, and many who came hoping for quick success left after only a few seasons, discouraged by the heat, aridity, and distance from polite society.

Land speculators had already entered the valley before the railroad, and some of their early investments turned out well. In 1868 San Francisco capitalists William S. Chapman and Isaac Friedlander persuaded a group of largely German investors to buy newly available land-grant college scrip[1] and use it to claim an 80,000 acre block of land in Fresno County some four years before the railroad reached what would be Fresno station. This so-called German Syndicate consisted of 21 investors who joined together to promote settlement on land between the Kings and San Joaquin rivers. In 1873, however, the syndicate dissolved just as the railroad opened its land office, and some of the individual investors began subdividing their holdings into small-parcel colonies based on ethnic or religious affiliations. Colony promoters also built roads and irrigation ditches, set out trees and vines, and offered advice to those who lacked farming experience. The first such colony was the Central Colony, established in 1874. Within a decade some twenty colonies covered 45,000 acres, providing homes for 1,500 families in the Fresno area.[2]

Henry Miller was aware of the gathering wave of land development and was not one to let opportunities slip through his fingers.

Henry Miller (top) and Charles Lux were recognized for their success when their portraits were included among the prominent Californians featured in a series entitled "Men We Know" that appeared in San Francisco about 1880.

Courtesy of San Francisco History Center, San Francisco Public Library.

Squatters near the Santa Rita Ranch, who had no title and were paying no rent, were already beginning to raise wheat on the open ground and even to practice some primitive irrigation. To forestall a greater influx of settlers, Miller quickly bought up portions of sixteen different land grants scattered in several counties. By 1870 the two partners were in possession of more than a hundred thousand acres in the San Joaquin Valley and nearby foothills, and by 1872 it was reported that there were 68 miles of fence around the Miller & Lux properties, five boards high, extending from Hill's Ferry on the north to Firebaugh on the south. By 1880 the firm not only owned a fifty-mile-long stretch along the San Joaquin River but other big tracts strategically chosen along several waterways—an astonishing 600,000 acres sprawling across half a dozen counties, all of it purchased at prices that would later double and quadruple in value.

Throughout the 1870s the Miller & Lux partnership was a juggernaut in land acquisition, propelled by Miller's relentless explorations and financed through Lux's skill at raising capital. Lux would have liked to invest in stocks and bonds, residential lots and other urban opportunities, but Miller concentrated on rural land and the cattle business, and his views prevailed. During these years Miller moved around mainly on horseback, constantly scouting possibilities, and his powers of observation were keen. He noted the lay of the land, the waterways, and the patterns of vegetation, and considered how these might support livestock; he noted the weather and the wildlife, and calculated how they might relate to large-scale ranching throughout the year. From water supply to soil characteristics to micro-climates created by topography, he wanted a buffer against the environmental extremes encountered in the previous decade, and he filed the details away in his retentive memory. Uninterested in wheat as a speculative crop and largely indifferent to fancy livestock breeding, Miller viewed the partnership as a purveyor of meat from start to finish, and to serve that goal the firm needed extensive and varied tracts of ground to juggle seasonal needs.

Always on the lookout, Henry Miller was ready at any time to purchase acreage from those depressed by hardship or lured by other opportunities, but always at his own considered price. He did not bargain or haggle, but generally named a sum he would pay, and if the seller would not take it, there was no sale. For a few years William S. Chapman acted as an agent for him in acquiring land scrip, but Miller himself selected the properties the firm bought and generally got them at low prices when land was cheap because it was considered unprofitable or because the owners were under financial pressure. Usually Miller would also buy the livestock on the land, the sale of which might sometimes cover the costs of the purchase.

Though the partners never carried mortgages on their own land, they sometimes extended mortgages to other landowners, giving them reasonable time to redeem the loans but foreclosing when the time ran out. Miller bought homesteads and preemptions from small farmers and bought big ranches that had gone belly up. Year by year, acre by acre, he accumulated land, and he and his agents became familiar figures in county courthouses. Like several other cattlemen, he made entries under various laxly administered land laws: soldiers' and school warrants, Swamp and Overflowed Lands laws, the Timber Culture Act of 1873, the Desert Lands Act of 1877. One of his employees later remarked that Miller might have "cut a few sharp corners" in buying land, but that in those days the corners were so wide it was sometimes hard to tell if they were square or round. It was generally believed that Miller drove hard bargains and took advantage of opportunities, but never cheated or engaged in fraud.

One famous story that circulated about a Miller method of acquiring land may have been apocryphal (for the tale was told about others as well), but it was told and retold until it became folklore, and some local residents swore it was true. It had to do with swampland, which could be claimed from the government for a relative pittance if the claimant agreed to drain it. Land was defined as swampland if it was navigable by boat during high water periods. Miller, it was said, swore that he had traversed certain properties by boat, but the boat in question was actually mounted on a horse-drawn wagon. Miller might, in fact, have followed the practice of some farmers in areas prone to flooding who had designed wagon-boat combinations that could float through submerged swales if needed with their wheels in the water. True or not, the story hinged on the bending of regulations that Miller might well have found ripe for interpretation.

His philosophy was always to buy, never to sell, for he believed that in the long run land was the basis for all economic gain. "The amount of land is limited and the population is unlimited," he told an interviewer. "I think, good land, there is no telling how high it will go. There is a time coming when people will find it out. I think there is no doubt but that California will be the most densely populated of any portion of the United States." [3] At its peak Miller & Lux owned 1.4 million acres of land in three states and controlled nearly ten times that amount, and only reluctantly did Miller let go of any property.

Henry Miller was the epitome of the traditional German virtue of thrift. He had learned young how to get the most from modest means and had seen for himself that frugality and efficiency could be the basis for prosperity. Throughout the years he closely oversaw all the Miller & Lux properties, traveling constantly from ranch to ranch inspecting facilities, ordering supplies, and giving directions on every aspect of daily life from the management of animals in the pastures to the disposal of scraps from the kitchen. In the risks he undertook he remained conservative, taking action only after considered judgment and insisting on personal control. Believing that speculative enterprises were foolish, he learned from experience and applied what he learned toward the improvement of business. He had seen the devastation caused by the drought of 1863–64 and never forgot the lesson. As fast as land was acquired, he had dams and levees constructed along the waterways so that the ranches could flood irrigate the fields used for pasture or feed crops. "In California every year is a dry year until it proves itself otherwise," he wrote to his ranch managers, and directed that they should always grow surplus quantities of hay to stockpile in case of need. That policy paid off. During the dry years of the 1870s many California cattle shrank down to skin and bones, but not those of Miller & Lux. The partners not only maintained their herds but were able to acquire more land when other ranchers went bankrupt. In later years when harsh winters broke the ranchers of high-country Nevada and Oregon, Miller's prudence paid off there too.

Miller was constantly on the move. Said to be a fine horseman as a young man, he rode only the best horses and spent many hours in the saddle. As he grew older he began to use a buggy, with or without an employee driver. While Bloomfield Farm near Gilroy served as his home base, the Canal Farm near Los Banos became the company's headquarters in the San Joaquin Valley. Miller frequently stayed at the large ranch house built there for employees, but every Miller & Lux ranch kept a room available for him at a moment's notice. Wherever he showed up his arrival set off tremors, for he went through the ranches like a whirlwind—fields, barns, corrals and feedlots, bunkhouses, dining halls, and kitchens—making mental notes and giving non-stop directions. Only occasionally did he check into hotels.

Almost every day he wrote letters: to Lux in San Francisco, to ranch supervisors, or to others in the trade–thousands of letters over his lifetime, almost all strictly about business–reports of buying and shipping cattle, hogs, and grain; descriptions of the weather; orders for lumber, feed, and other supplies; directions for foremen; needs for employees to go places; needs for new employees to be hired, including housekeepers and cooks. English was Miller's second language, and although he wrote colloquially he was never perfectly fluent. Writing in a hurry, he often misspelled simple words like "atend" and "verry." His letters were usually scrawled on small tablets of ruled paper, though he sometimes sent telegrams or used hotel stationery.

He was a good judge of men–and of women too, though there were few women on the ranches. He chose employees with care, expecting them to work hard and requiring absolute obedience to his orders. Though he was a demanding taskmaster, he earned the respect of many loyal workers who stayed with him for years, finding him a fair boss who listened to what they had to say and was always interested in suggestions. He insisted on good food and accommodations for his employees and was generous in times of need and at holidays. In some ways his relations with underlings were almost feudal, yet he got along well with most of them. Charles Nordhoff, staying in a San Joaquin Valley hotel one evening, observed Miller interacting with his men there: "It happened that Mr. Miller was present that evening. He is a German by birth, short, alert in his movements, neatly dressed, and with a good nose, a low forehead, and a remarkably expressive countenance. He busied himself with instructions to his foremen, smoking meanwhile, and I was amused to see that when once he got up from his chair, a vaquero slipped into it—and as there were more people than chairs, after wandering about the room for a while, exposed to a battery of tobacco juice, the owner of half a million acres finally squatted down in the wood-box at the side of the fire, and prudently took care not to get up again until he went off to bed....His men give him a good name for liberality and kindness to them, and to all working-men; and I believe there is no doubt that he is an uncommonly able business man." [4]

The lifeblood of the ranches was water, and no one knew that better than Henry Miller. He prized the riparian areas of the properties he acquired, studying how best to control and impound water when it was seasonally available and how to release it later when it was needed. Like the careful husbandman he was, he gave detailed instructions on using water efficiently. He always assessed the potential for competition for limited supplies and from the beginning of his long career as a land buyer kept careful records of water rights. Those records not infrequently became the basis for litigation with other users over the years.

Miller's influence extended far beyond the firm's ranches, however, for he helped shape the water history of the entire San Joaquin Valley. Within the context of his own interest, he oversaw the building of canals through Miller & Lux land, bought up and restored a failing canal company, and fought long battles in court over issues that would plague state water law well into the future.

Miller & Lux began to get involved in water activities in a big way during the early 1870s. California's wet winter of 1869 was succeeded by a stressful two-year drought in 1870–71, which limited wheat growing and discouraged livestock men in the San Joaquin Valley. Because of the company policy of stockpiling hay, Miller & Lux survived better than most others, but their cattle suffered too, and Miller arranged for some herds to be driven north to the remaining damp spots in the tule lands of the Delta. Meanwhile, it became ever more apparent to the San Francisco capitalists who aspired to make money on valley land that the future of their investments would depend on the availability of water, not as nature capriciously provided but as engineers and developers pragmatically planned.

In the booming late 1860s San Francisco entrepreneur John Bensley had already begun a vigorous effort to find a way to make the dry plains of the San Joaquin bloom in summer as they did in spring. Bensley had a successful record in mining, shipping, and water development for the city of San Francisco; now he set his sights on the interior valley. Under the current laws governing water rights, Bensley and a partner, Orrin Simmons, posted a notice at the confluence of the San Joaquin River and Fresno Slough stating that they would be appropriating water through a canal built for the purposes of irrigation and transportation. They then planned out in general terms an ambitious canal system running all the way from Tulare Lake in the south to the Delta in the north—a distance of some 160 miles—to divert water onto adjoining lands. Financially, however, this project was far beyond their means, and after an abortive attempt to begin canal work, Bensley offered Miller & Lux a chance to invest in it in 1870. Though the partners were interested in the idea, they were also skeptical about the scope and scale of the project, as well as Bensley's ability to carry through, and they declined the opportunity.

Bensley then used his connections to attract some of the San Francisco capitalists known to be interested in land development. Wheat broker Isaac Friedlander and banker William Chapman, who already owned thousands of acres of valley land, enthusiastically took up the project in 1871. They reorganized Bensley's original company into the San Joaquin and Kings River Canal and Irrigation Company (SJ&KRC&ICo, later dubbed the company with many initials), offering investors a total of 100,000 shares to raise one million dollars for construction of an engineered waterway extending from Tulare Lake to San Francisco Bay. The project would be designed not only to provide water for irrigation but to serve as a transportation corridor for steamboats. The ten major stockholders included Friedlander, Chapman, W. C. Ralston of the Bank of California, Wells Fargo president Lloyd Tevis, and other moneyed men with a taste for grand ideas. One historian has described the initial investment scheme as a mixture of "giddy optimism with chicanery," and certainly the primary backers minimized the potential problems of canal construction in the far-flung valley while painting in deceptively glowing colors the potential for making profits from land development along the corridor.[5]

Friedlander and Ralston persuaded the new canal company to engage the well-known British engineer Robert Maitland Brereton, who had recently completed work on the Grand Indian Peninsular Railway from Bombay to Calcutta, to prepare a feasibility study on developing the valley's water resources. Looking for new challenges, Brereton was glad to oblige. After a six-week survey he reported back with recommendations for *two* major canals, one tapping the San Joaquin and Kings rivers and the other, much bigger, leading from Tulare Lake. The report also recommended a complex of smaller canals to be constructed throughout the valley in ongoing stages.

Brereton himself was a practiced promoter. After submitting his report he began to lobby in Washington, D.C., for federal canal development funds and also sought backing in England for colonization efforts like those in Fresno County. He produced an attractive pamphlet called "Propositions for English Upper Class Colonies in California," published in June 1875 in London, as well as another that touted the potential for prosperous small farms for middle-class and yeoman settlers. These pamphlets urged affluent English families to form cooperative associations to buy tracts of land in California, suggesting tactfully that younger sons and nephews not in line for inheritance might be sent abroad to oversee the family investments—and, of course, those investments might well double in value after irrigation became available.

Miller still had serious doubts about the grandiose scope of the project and the ability of the canal company to achieve its goals, but when Bensley approached him again he decided to take a limited number of shares to keep abreast of developments. He could certainly see the advantages in being part of the project, for much of his land lay on the valley's dry west side. He proceeded to protect the firm's interests by negotiating a set of detailed contracts with the canal company. Miller & Lux would grant the company a 50-mile right of way across lands on the west side of the San Joaquin River, plus the use of sloughs and the willow brush along the river, a $20,000 subsidy; and use of Miller & Lux employees and equipment, in return for a fair share of the canal water at a reduced price. The canal company would provide irrigation water for about 50,000 acres of Miller & Lux land at a maximum annual charge of $1.25 per acre (half of what other water users would later pay) plus water for domestic purposes and use of "unused" flows at no charge. A separate agreement gave the canal company a five-year lease on a 6,000-acre tract of Miller & Lux land, on which it would lay out a system of ditches and plant alfalfa to demonstrate the value of irrigation. The tract would serve as a "model farm" to show others.

Paperwork was one thing; supervision was another. Miller was not one to leave construction activities on his property in other hands. He himself supervised the canal work for the fifty miles through the firm's land. For months he oversaw the job of digging the canal as well as the installation of a surprisingly effective brush and earth dam 350 feet long across the San Joaquin River. He also had his men build hundreds of miles of lateral ditches, laying out a system that would efficiently water Miller & Lux fields when the ditch gates were opened.

The 78-mile-long SJ&KR Canal, constructed from 1871 to 1873, began at a point six miles south of the Firebaugh Ferry in Fresno County, angling northwest from Fresno Slough through Merced County and almost parallel with the San Joaquin River into southern Stanislaus County. The canal was designed to be 68 feet wide at the top, tapering to 45 feet at the bottom for the first 38 miles, then shifted to 53 and 35 feet respectively for the remaining length. Construction was an arduous undertaking, taking its toll on the hundreds of men and animals working with the fairly primitive slip scrapers of the time, but when things went well, the crews could make progress at the rate of two-thirds of a mile a day. Overall, however, the enterprise turned out to be a costly proposition, for since no railroad yet entered the area, all supplies–machinery, lumber, and other freight, including food and feed–were brought either 150 miles up the San Joaquin River by barge during high water season or 50 miles over Pacheco Pass from Gilroy by wagons. Construction costs soared to well over a million dollars. Miller said later that the building of the SJ&KR Canal cost twice what it should have.

Contemporary writers expressed great expectations for the canal, not only for its potential to irrigate nearly three million acres ("...every acre of which, with water, could produce two crops a year"), but also for its role in the reclamation of a million acres of swamp and overflow land. Would-be settlers were at first enthusiastic about the prospects for subdivision of large tracts of land into small farms with the promise of intensive agricultural development. When the canal began flowing, however, the company charged $2.50 an acre for water delivery, a cost that water users found outrageous when they saw what they regarded as inefficiency in canal operations. Moreover, big landholders like Miller & Lux made no move to partition their acreage to sell to newcomers, only continuing to acquire more land themselves. By 1874 public opinion was beginning to turn against the canal company. At the same time company stockholders were growing restive as they continued to pay assessments for construction and management while receiving few of the promised dividends.

In the fall of 1874 Henry Miller was seeing a doctor and taking medicine for an inflamed throat. For more than twenty years he had worked with little respite, and completion of the canal through Miller & Lux land brought him to the point of exhaustion. The following spring he travelled to Germany for medical treatment, spending seven months there visiting relatives and recuperating. In his old village of Brackenheim he put up a memorial stone to honor his deceased parents and arranged for his sister and other kin to come to California.

He had much to catch up on when he got back to business. The year 1875 was a difficult one, with the lingering national economic depression exacerbating California's own woes. Bankruptcy haunted Isaac Friedlander as well as other San Francisco investors. Following risky speculations in the Comstock Lode and elsewhere, the prestigious Bank of California nearly collapsed, and the temporary closing of its doors was followed by the startling death of its popular president William C. Ralston. Under these trying circumstances the San Joaquin and Kings River Canal Company also began to falter, a victim of escalating costs and poor management. While Brereton and other company officials received inflated salaries, ditch managers bungled seasonal procedures. The company spent large sums of money trying unsuccessfully to get federal or state subsidies, and the stockholders quarreled. Brereton proposed a reorganization, but after his efforts to raise money in England failed, he finally left California behind to move on to other projects.

Not surprisingly, when canal officials tried to get more money from Miller & Lux, the soured relationship led to a lawsuit over the original contracts. Following operating disagreements, the SJ&KRC&ICo began to withhold water from Miller & Lux lands. When Miller returned from abroad he is said to have become so incensed at what was going on that he personally took an axe to the canal gates along the firm's levees, directing his employees to distribute water out into the fields. With a warrant out for his arrest, he and Lux began contacting other stockholders and arranged to buy up enough stock from disgruntled investors to give the partners a controlling interest in the company. They got a considerable number of shares at reduced prices, some of them from Brereton and some from the promoter John Bensley, who had gone bankrupt. At the company's annual meeting in 1877 Miller & Lux reported in with just over a majority of all the stock, acquired at less than one-third of the original investment, and Charles Lux was seated on the board of trustees. The partners could now control canal operations and stamp their own style of management on the company. The charges filed against Miller were dropped.

Miller & Lux now virtually owned the first large-scale canal system in the San Joaquin Valley. Taking charge, Miller made sure the big canal was in good condition and efficiently run. He ousted the preceding officers and brought in new staff for less money; installed John Q. Drummond, who had previously supervised canal construction, as the competent new general superintendent; and moved the company offices from San Francisco to Center Point. Insisting on more reliability and economy in operations, he also directed that per-acre charges for water be made appropriate to usage and more reasonable in price—making sure, of course, that Miller & Lux's own rates remained minimal.

The SJ&KR Canal became known as one of the best-run and cheapest water systems in the state. During the next few years Miller supervised the construction of hundreds of miles of new lateral ditches. In 1878, at settlers' request, he extended the canal from Los Banos to Crow Creek, and shortly afterwards began construction of the Dos Palos and Temple Slough canals. By 1880 nearly all Miller & Lux lands were watered from a complex of canals that permitted close monitoring of fields, and other settlers began to raise irrigated crops like alfalfa, encouraging dairymen to begin moving into the area. Thus some 150,000 acres of dry land in the San Joaquin Valley became more productive, and the value of the land increased. Henry Miller in his careful and practical way had improved not only his own property but that of many others. He had helped bring the San Joaquin into a new era.[6]

Land monopoly in California was a major issue in state politics by the end of the 1870s. Homesteading in other parts of the West had helped develop many new communities, but in California the "curse of land monopoly" stymied successful rural development in some parts of the Central Valley as long as speculating capitalists or holding companies controlled vast tracts. It was because of his reputation as a land baron that Hugh Glenn became the target of ferocious attacks by the Workingmen's Party and was defeated in the 1879 election. During that campaign Denis Kearney and his colleagues spent time in the San Joaquin Valley rousing high feeling among the squatters frustrated by the problems of obtaining land titles. Hopes that land monopolies might dissolve with the influx of settlers brought in by the new rail lines—and the advent of irrigation through the new SJ&KR Canal—had not yet materialized. Capitalists with deep pockets still retained advantage.

Miller and Lux had deep pockets too, and they had little intention of selling what they had accumulated. But the feeling among their neighbors was deeply ambivalent. Some resented the partners' huge holdings; others admired their well managed ranches. Miller &

Lux ranches were productive, provided jobs for hundreds of people, and showed others what was possible.

Miller himself escaped some of the squatters' hostility because he was not a speculator. He bought and developed land for ranching, often to the benefit of the neighborhood, and he seemed to be committed in the long haul to improving the valley's potential. It helped considerably that Miller himself, though by now wealthy, was clearly a working man, lacking in pretensions and modest in his habits. He took pains to be on good terms with his neighbors and with laboring men in general, believing that "A friendly neighbor is a great asset," as he often reminded his ranch managers. His regular physical presence in the valley and his judicious but genuine generosity to local residents helped make him a figure to be respected rather than despised. A body of popular anecdotes, some humorous but none of them ugly, began to collect around his personality.

The resentment against land monopoly in general, however, was graphically demonstrated in May 1880 with the Mussel Slough incident in Tulare County—a harsh culmination of escalating local tensions. By the early 1870s the Central Pacific had laid hundreds of miles of track in the San Joaquin Valley, and immigrants hoping to acquire land arrived in growing numbers. Hundreds squatted along the projected rail line in the Mussel Slough area south of Fresno, expecting to be able to buy parcels in the railroad sections at low prices. These hopes were dashed by route changes and announced price hikes. Squatter hostility simmered while matters were being clarified in court, but finally it broke into a boil on a farm near Grangeville. In an armed confrontation between railroad agents and locals angered by rumors of eviction, gunfire erupted and several men were killed.

A federal grand jury subsequently indicted 17 men and found five guilty of interfering with a marshal in the performance of his duties. The five went to prison, but public sentiment was in their favor and they were widely regarded as heroes defending their rights. After a few months the prisoners were released, but the episode was sobering for everyone. Ultimately, the railroad company made small reductions in its land prices, and most of the squatters ended up buying the land on which they lived. The Mussel Slough story became part of California history (and folklore) and a generation later was reinterpreted and dramatized by Frank Norris in his novel *The Octopus*.[7]

The decade of the 1870s saw a marked decline in California's original range-fed cattle industry. Less free rangeland was now available to support independent cattlemen as public lands passed into private ownership and grazing areas were put to the plow. Most of the remaining unoccupied land with water holes was now controlled by a few large companies like Miller & Lux, and these companies were beginning to raise forage crops to supplement range feeding. Between 1870 and 1880 the number of cattle in California actually dropped as the business changed.[8]

There were several reasons for the decline of range feeding in California. The passage of the "No Fence" law in 1872 repealed the Trespass Act of 1850, which had required farmers to protect their planted fields with a fence four and a half feet high of stone, or five and a half of lumber, if they wanted payment for damage done to their crops by free-ranging cattle. The great burden of expense thus imposed on the farmer clearly favored the freedom of stockmen to let their animals roam. The new no-fence law, however, required the stockman to fence his animals in, rather than the farmer to fence them out. This law did not apply to the mountain counties or to the southern San Joaquin Valley until after 1874 because of the enormous cost of fencing large tracts there. The law encouraged a large increase in barbed wire patents taken out during the 1870s, however, making fencing more feasible. Gradually, no-fence laws extended through the entire state. Miller & Lux had fenced most of their holdings since the 1860s, but other ranchers now had to comply with the new conditions, go out of business, or move to other states.

Rangeland resources had deteriorated over the years as well. During the Mexican era and the early years of American livestock production, the native perennial grasses of California–primarily bunchgrasses of the *Stipa* and *Poa* and *Fescue* genera–were plentiful and of exceptional pasturage quality. Some of the native herbage was also highly nutritious. Livestock men took full advantage of the free feed, and the rangeland in many areas became overrun with their herds–especially large bands of sheep, the "hoofed locusts" so detested by John Muir. Overgrazing had a deleterious effect. The original bunchgrasses were eaten to the ground or scuffed out of the soil by animals' hooves, while new annual foreign plant species came in uninvited, transferred with human activity. Some new species were of good forage value, like wild oats and bur clover, but inferior plants like pigweed and cheatgrass began to replace the natural grasses. Drought years hastened the deterioration, until only in the back country were bunchgrasses still predominant. Eventually, like Miller & Lux, cattlemen began to counter recurring drought cycles and lean winters by producing their own hay, and stock farming began to replace most of the former range feeding.

Disease made inroads too. "Texas fever" became a scourge in California by the late 1870s, killing large numbers of cattle before it was discovered that cattle ticks were responsible for the spread of the disease. It was not until the 1890s that government scientists were able to help ranchers manage their grazing grounds to avoid the spread of the disease, and to develop vaccines for other maladies like "blackleg" and anthrax.

Transportation developments changed cattle marketing. Soon after the railroad was extended from San Jose to Gilroy (1869), for example, Miller's Bloomfield Farm opened its own station to load livestock into rail cars bound for the city, ending the drives of cattle that had earlier clogged the roads. Railroad construction in the San Joaquin Valley soon offered a more convenient option for shipping cattle from points south too. The long trek over Pacheco Pass also became easier as roads were improved. In 1879 the two counties of Santa Clara and Merced bought out the toll road company and opened that road as a free public thoroughfare, meanwhile surveying a less strenuous alternative route higher on the hills. The "New Grade" road, completed in 1885, offered travelers far greater accessibility and comfort, but with growing traffic the historic big cattle drives over the pass continued only a few years longer.

Many small free-range cattle ranchers in California were driven out of business by the continuing loss of public rangeland, deterioration of resources, and new disease. Nevertheless, most cattlemen adapted to the new times, and the number of cattle in the state rebounded as stock farming grew. Like Miller & Lux, ranchers Richard Gird and Edward Beale in Southern California built up great spreads too, acquiring property through the land laws, focusing on water rights, and paying attention to the details.

California's free-range cattle industry declined almost a decade earlier than elsewhere in the West. Cattle kings and cowboys reigned on the Great Plains during the same years that saw the rise of Miller & Lux in California, but the evolution of the livestock business there was different. From Texas northward the great continental cattle drives began in the 1850s, supplying meat to emigrants along the Oregon Trail and to mining towns in the mountains, with the trade increasing after the Civil War. Cattlemen prospered as the vast herds of native buffalo and the Indian way of life vanished. The Great Plains saw an orgy of slaughter by reckless American and

European big game hunters—some even shooting for sport from moving trains. Millions of buffalo disappeared from the plains not long after the Civil War, memorialized by great piles of bleached bones. Market hunters completed the slaughter in outlying areas, and cattlemen increasingly held sway over the country. The herds that then darkened the plains were free-range cattle, until severe weather and tough economics forced a change.[9]

Until land-seeking farmers began to try to wrest a living from the stubborn plains, cattlemen ran their animals over the free range to several market towns, subsidized by wealthy investors who gambled on favorable beef prices in an expanding American economy. The ready flow of investment capital, often from abroad, along with the rise of urban markets, provided the basis for the growth of cattle herds and shipping centers in mid-continent. Large-scale cattle production was also encouraged by the extension of railroad lines and the opening of the public domain. In 1867 Abilene, Kansas, became the first of several railway cow towns from which cattle were shipped to Chicago.

The popular literature of the day endowed the Great Plains cattle drives with a special kind of romance. The cowboy working a herd along the trail became a frontier icon celebrated in western folklore—the tough, laconic horseman who could control stampeding cattle or sing songs by the evening campfire with equal flair. But speculation in beef prices was the reality, and cowboys were a minor part of the equation. Colorful characters who tried their hand at raising cattle in the West included Americans like the young Theodore Roosevelt in North Dakota, Charles Goodnight in the Panhandle, and Richard King in Texas, but Europeans moved in too, attracted by the adventure along with the heady prospect of profit in a growing market. Many British, a few French, and several Germans tried their hand at cattle ranching on the Great Plains.

The career of one of those Germans to some degree paralleled that of Henry Miller. Conrad Kohrs, an immigrant to America in 1854, became a major cattleman in Montana by the 1870s. A few years younger than Miller, Kohrs spent part of his youth as a cabin boy sailing the world and then in mining camps in California, Canada, and Idaho. He arrived in southwest Montana in 1862, where he found work as a butcher's assistant and then started his own shop in a gold camp. In 1866 Kohrs bought stockman Johnny Grant's cattle ranch in the Deer Lodge Valley. In time he acquired 30,000 acres and with his half-brother John Bielenberg ranged cattle over open country all the way from Colorado to the Canadian border, becoming known for quality. He helped found the Montana Stockgrowers Association in 1884. When the open range began to decline, Kohrs adapted to a mix of grazing and farming. He died in 1920, four years after Miller.[10]

The cattle kingdom on the Great Plains lasted barely two decades, from the end of the Civil War to about 1885. After the infamous killing winter of 1882 and the new range laws of the 1880s, the livestock bubble on the plains was largely over. Rangeland deterioration and another harsh winter in 1886–87 decimated the great herds and hastened the change from open range to enclosure ranching. With the days of the great cattle drives over, the independent stockman found himself more and more tied to a centralized marketing system based in Chicago. Many of the speculators were forced to sell out.

The cattle industry of California was similar in some respects to that of the Great Plains, yet had its own characteristics. Cattle succeeded elk and pronghorn in California as they followed the buffalo on the Great Plains, but the wildlife slaughter was never quite as horrific in California, and the cattle industry itself, with more limited markets, was never quite as speculative. California weather could be bad, but never as harsh as inland. Although many independent free-range cattlemen were squeezed out, other California stockmen built up their business through the steady acquisition of their own land. Miller & Lux almost from the start raised cattle in a tightly managed system that ended in the firm's own butchering facilities in San Francisco, moving the stock from field and feedlot to slaughterhouse and sales. Long after other cattlemen had vanished, Miller & Lux remained.

The hub of the Great plains cattle industry nearly from its beginning was in Chicago, that muscular city on the Great Lakes that became the brokerage point between the American East and West and a symbol of nineteenth-century enterprise.[11] The sprawling Union Stock Yard that opened on Christmas Day in 1865 on the south side of the city became the destination point for untold numbers of livestock over the years. The Stock Yard was a half mile square, an enormous complex of pens and sheds designed inside a broad loop of rail lines that made shipping thousands of animals in and out highly efficient. Slaughterhouses surrounded the central maze of corrals and passageways, and a large hotel and an exchange building provided facilities for traders as they connected by telegraph with markets across the country. Few who worked in or visited the site would forget the stench or the gore in the slaughter yards.

Chicago meat packers and distributors came to exercise enormous power over ranchers on the prairies and middlemen in the small towns, and radically changed the nature of the cattle trade as the United States grew. Two dominant figures in the Chicago trade transformed the industry during the years that the range cattle business dominated the Great Plains. Gustavus Swift and Philip Armour established themselves near the Chicago Union Stock Yard in the mid-1870s.

Swift began his career as a shrewd young farmer's son in Massachusetts who started his own butchering business while still in his teens. He moved west as the cattle markets shifted, eventually settling in Chicago, where he first sent live cattle east and then began experimenting with shipping dressed beef in rail cars. His partnership with his brother was incorporated as Swift & Co. in 1885, and Swift began to revolutionize meat processing. He had an engineer design an effective ice-cooled car for rail shipments, and the company soon financed the manufacturing of refrigerator cars. Swift & Co. built meat packing plants in several other Midwestern cities as well as in Texas, and shipments of dressed beef expanded nationwide during the 1880s.

Philip Danforth Armour was Swift's chief rival, with an interest in grain dealing as well as in meat packing. Armour had earned a stake in the gold fields of California by the time he was 24 and used those funds to get into the meat business, first in Placerville, California and then in Milwaukee, Wisconsin. In 1867 he founded Armour and Company, a food processing and chemical manufacturing firm headquartered in Chicago. Armour also developed a refrigerator car line and pioneered new principles in meat packing, including the use and sale of many former waste products such as hair, blood, bones, and cartilage.

By the late 1880s Chicago controlled much of the nation's meat supply in the Midwest and along most of the East Coast, and the two great meat packing companies' ruthless cost-cutting began to put competitors out of business. In the Far West the cattle market operated independently for some years longer, but not without the threat of a takeover by the Chicago packers in the 1890s—a story told in the next chapter.

By the romantic reckoning of "the old west," butchers and meat packers were tradesmen toiling in slaughterhouses and offices, while cattlemen were lords of the great outdoors. Critics often called Henry Miller a "butcher" instead of a cattleman. The truth is that he was both. He knew cattle and how to raise them; he knew herding and the work in the corrals; he knew his vaqueros and valued them highly. He intimately understood the land and water resources on all the company ranches. But cattle were raised for beef, and beef was a commodity, and markets were always the primary end. Miller could take pleasure in the sight of sleek cattle on green fields, but he could also estimate almost to the pound the amount of dressed beef that a grazing animal might yield.

A primary reason for Miller & Lux success was that the firm ran a highly controlled operation from rangeland and pasture directly into city markets, backed up by financial and legal expertise—what today is called a vertically integrated business. Though Miller himself was usually on the road visiting ranches and observing conditions in the countryside, the Miller & Lux corporate offices were sited on California Street in downtown San Francisco and in hard-working Butchertown near Hunter's Point.

Henry Miller's first slaughter yard had been located in the San Francisco neighborhood that became Chinatown, but was soon forced by city growth to move to Fifth and Howard, south of Market Street. At that time there were several slaughterhouses selling wholesale meat to many small retail markets in the city and surrounding area, but in the 1860s growing city pressures and objections to disagreeable odors made it necessary for all these butchering businesses to move out of town. The new Mission Bay or Long Bridge had just opened up what was then South San Francisco along the bay shore—the Potrero district or what is now known as Bayview—and this area immediately south of the city became known as "Butchertown."

The site was officially established in 1868 by an act of the Legislature after the butchers petitioned their representatives to obtain a permanent location, preferably over tidewater to dispose of waste. The legislation granted the butchers a tract of land along the south side of Islais Creek, a tidal channel. The tract was formally named Butchers Reservation and immediately subdivided into lots to be leased for 99 years. Led by Miller & Lux, several butchering firms built a long, broad wharf on pilings over the water to take advantage of tidal movement to disperse leftovers from slaughter. The wharf was big enough to hold a dozen or more slaughterhouses, and Miller & Lux built two of them. These became the final destination for livestock coming into the city.

Butchertown lay within three miles of the city center and within a few years became accessible by railroad from the south. Its core was today's Third Avenue, formerly Railroad Avenue, where trains could roll in to disgorge their bawling loads to vaqueros, who herded them into corrals at Hunter's Point. The long wharf extending into the bay also allowed big drover ships to unload cattle after dark, when the animals were less likely to be terrified. Cattle were kept in confinement up to three days before being killed to allow them to cool off from the trauma of travel.

By the early 1870s there were 16 slaughterhouses in Butchertown, where according to John Hittell some 300 cattle, 500 hogs, and 2,000 sheep were prepared daily for the markets. The slaughterhouses utilized everything–blood, fat, hides, hair, tails, horns, and feet–selling these byproducts for use in a variety of manufactured goods. Twice a day the slaughterhouses were scrubbed out to keep the premises clean.

Butchertown was a tough and odorous place, smelling of blood, sweat, and offal even from a distance. Many slaughterhouse employees were Germans who like Henry Miller himself had learned their trade in the old country; others were of varying origins. Affluent San Franciscans tended to look down on the workers of Butchertown, but they formed a proud and cohesive community among themselves. The butchers were well paid and powerful men, often tall and bulky, sure of their strength and skills. It took strong arms and backs to stun large animals with sledge hammer blows to the head, then hoist the bodies onto overhead tracks where throats were slit and the carcasses rapidly dismembered. After hard workdays Butchertown workers often consumed vast quantities of beer, and a favorite local treat was the shrimp caught off Hunter's Point by Chinese fishermen.

By 1880 Miller & Lux employed more than 20 men in the firm's slaughterhouses. The firm slaughtered some 83,300 animals that year, including steers, cows, calves, hogs, sheep and lambs. In 1881 Miller & Lux was said to have sold $700,000 aggregate value of livestock in the San Francisco market. A few years later the *San Francisco Call* estimated that the firm supplied an average of 52 steers, 70 hogs, and 200 sheep daily to city meat markets.

As on the ranches, Miller insisted on both efficiency and good sanitation practices in the slaughterhouses, and he encouraged cooperation among the butchers to help keep meat prices stable. Prices at Butchertown were more or less dependent on those set by Miller & Lux, the largest firm. When others urged him to raise prices, Miller tended to resist, refusing to gouge retailers when cattle supplies were tight and aiming, he said, to keep the market steady for everyone's benefit. Over a long period of time Miller was credited for his leadership in keeping quality high and prices reasonable.

The original Butchertown slaughterhouses lasted for nearly 40 years, until the great San Francisco earthquake and fire of 1906 destroyed them. Afterwards they were rebuilt as more modern "abattoirs." Butchertown prepared meat for the San Francisco and Bay Area wholesale and retail markets for nearly a century, until the 99-year leases expired in the 1960s.[12]

The wealthy citizens of San Francisco built elaborate houses from the 1870s onward–highly decorated wooden gingerbread mansions or imposing stone palaces–usually sited high on the hills to take advantage of the splendid views. Most of these mansions belonged to financiers, mining kings, railroad magnates, or prosperous merchants, but the "cattle kings" soon acquired big houses too.

Though they always maintained rural residences, their primary family homes were in the city.

In the 1860s Charles Lux and his family moved into the fashionable South Park neighborhood and lived there until industry started moving in nearby. About 1878 Lux had a luxurious new home built at 1900 Jackson on the corner of Gough in Pacific Heights, a site with a magnificent view of San Francisco Bay. He also maintained a well-appointed house on his estate at Baden south of the city, surrounded by gardens besides the barns and corrals. Miranda Lux began to pursue philanthropic interests, taking an interest in education and vocational training for working-class girls and establishing a free kindergarten for needy children. Moving in elegant social circles, she became friends with other prominent women like Jane Stanford and Phoebe Hearst, but to Miller's disgust she never took an interest in the business of the partnership.

Lux had no children of his own. His stepson Jesse Potter, Miranda's son from her first marriage, went to a private school in Benicia and grew up in luxury. Ever the critic of pampered youth, Miller disapproved of Jesse's influence over his own son, Henry Miller Jr., who also developed the habits of the rich and careless as he grew to adulthood.

Another fashionable San Francisco neighborhood in the 1860s was Rincon Hill south of Market Street, rising up from Rincon Point at the northeast corner of the peninsula and overlooking the bay toward Oakland. Unfortunately for homeowners there, the city allowed real estate promoters to gouge an enormous chunk out of the hill in 1869 to expedite travel. A chasm was cut up to 75 feet deep from Folsom to Bryant streets, and the earth taken out of the infamous Second Street Cut was used to fill South Beach cove at the bottom of the hill. Boosters of the cut said it would help move traffic between Market Street and the main waterfront and the growing industrial area to the south, including Butchertown and the new

Pacific Mail Docks, while providing a better grade for the incoming rail line. The earthmoving, however, proved disastrous. Landslides took at least one mansion downhill, and other homes sustained serious damage. Although the hill continued to house polite society and attract a few celebrities like Robert Louis Stevenson, who lived there briefly, portions of the neighborhood began to decline by the mid-1870s, and working-class homes began creeping up the slopes.[13]

Perhaps because the price was right, Henry Miller bought property on Rincon Hill about 1875, at the northwest corner of Harrison and Essex streets. Miller had the original house on the property torn down and in 1877 completed a new and grander three-story wooden mansion with a mansard roof and a tower, where his family lived most of the time and he stayed while he was in town. The 1880 federal census listed the Miller family at this address in San Francisco, composed of Henry, a butcher, his wife Sarah, two daughters, one son, and three nieces, along with an Irish coachman and an Irish servant. Though many of his well-heeled neighbors eventually moved elsewhere, Miller maintained this city home for nearly three decades.

Bloomfield Farm, along the banks of Carnaderos Creek near Gilroy, was Miller's favorite residence. Here in the Santa Clara Valley the hillsides were studded with old oak trees, the pastures were green and lush, the climate salubrious. As Miller added acreage to what was called the "home farm," Bloomfield became a self-contained community, growing to include its own livery stables, blacksmith shop, poultry houses, granaries, general store, and railroad shipping station. Hundreds of employees and their family members lived on the ranch too. For years Miller himself lived for weeks at Bloomfield in relatively simple quarters when he was not traveling.

Not much is known of Sarah Wilmarth Sheldon, Miller's wife, a plain and down-to-earth woman devoted to her family. Married at age 20, she was 13 years younger than her husband and learned to live with his domineering ways. She gave Henry four children. Her first child died in infancy, but three survived: Henry, Jr., born in 1862, Nellie Sheldon, born in 1865, and Sarah Alice, born in 1871. The family lived primarily in San Francisco or for periods at Bloomfield, sometimes visiting the Santa Rita Ranch. Miller traveled so much that he was seldom at home with his family, though he usually tried to be there on Sundays.

His relationship with his children was typically autocratic, a legacy of his own German childhood. Though he valued the tradition of family life, he hardly knew how to express affection, and at the dinner table the young ones learned not to speak unless they were spoken to. His only son Henry, Jr., was frail as a child, and father and son were never close. Later, as he became a young adult, Henry, Jr,. was reported to suffer from a variety of ailments, some of them apparently brought on by dissipation. Miller's daughter Nellie, on the other hand, was a sociable child who married young but remained devoted to her parents.

His youngest child, Sarah Alice, was said to be her father's favorite and the most like him. In June 1879, however, Miller lost his lively eight-year-old daughter when she was thrown to the ground and killed instantly in a horseback riding accident near Bloomfield. She was buried beneath the cypress trees at Bloomfield near the graves of Miller's first wife and child. Miller never really got over the loss.

Depressed after his daughter's death and worn out by work and the heavy demands of litigation the following year, he began to suffer from what one of his housekeepers later called a nervous breakdown but a journalist reported as malarial fever. In April 1881 the *Merced Express* reported that "Mr. Henry Miller is very low, and little hopes for his recovery is (sic) entertained." In May the same newspaper reported that he was taking a trip to the eastern states

The San Francisco home of Henry Miller stood for nearly 30 years on the northwest corner of Harrison and Essex streets, until it burned down in the fire following the 1906 earthquake.
Courtesy of San Francisco History Center, San Francisco Public Library.

on the advice of his physician, who was accompanying him on the journey along with his son, a secretary, and a waiter. The *Gilroy Advocate* also noted his trip east, reporting that he was very weak but might extend his trip to Europe. Miller spent several months away from California recovering his health, among other places visiting New York's Saratoga Springs and Montreal, Canada.[14]

His condition much improved, Miller returned home to an enthusiastic welcome in Gilroy. He found himself, however, in the middle of one of California's landmark water litigation cases. Once again, he picked up the reins.

1 To raise money for developing land-grant colleges, the states issued scrip certificates allowing buyers to locate claims on public land.

2 Smith, *Garden of the Sun*, pp. 569–604.

3 Interview with E. Fowler reprinted in *One Man Show.*

4 Nordhoff, *California for Travellers and Settlers*, p. 192.

5 For the full story of the SJ&KRC&ICo, see Pisani, *From Family Farm to Agribusiness*, and *Water, Land, and Law in the West.* More details are included in *Engineers and Irrigation*, pp. 9–18, 32–33.

6 Sawyer, *One Man Show*, pp. 37–44.

7 See Beers, *Gunfight at Mussel Slough*, for a selection of the many publications inspired by the incident.

8 Pulling, *History of California's Range Cattle Industry*, follows the transformation from range to stock farm production.

9 Cronon, *Nature's Metropolis*, pp. 218–24.

10 National Park Service, "Grant-Kohrs Ranch."

11 Cronon, ibid., pp. 225–59.

12 Bowcock, *Butchertown: A Collage of a San Francisco Institution,* recounts local recollections of Butchertown.

13 Shumate, *Rincon Hill and South Park: San Francisco's Early Fashionable Neighborhood*, is a pictorial history of the neighborhood

14 The *Merced Express* reports of 1881 are reprinted in *One Man Show.*

~8~

Livestock Empire

IN THE ONGOING BATTLES over land and water rights that shaped San Joaquin Valley history and California law, Miller & Lux took up a prominent role, from their acquisition of the San Joaquin and Kings River Canal to aggressive defense of their holdings elsewhere by litigation over limited water supplies.[1]

The most widely publicized of their court cases was the Lux-Haggin lawsuit of the 1880s, which after several rounds of litigation became a benchmark in California water law. With two great landholders in an escalating confrontation over whose rights would take precedence, the case engaged many prominent attorneys' talents and attracted widespread attention. The problem in the courts at the time was that California law itself seemed ambiguous on the issue of water rights; some laws even appeared to be contradictory, and the inconsistencies had to be sorted out. Early courtroom interpretations of the state's water code led to appeals in which new court decisions overturned previous ones. It took a contentious five years to resolve this landmark case.[2]

The Lux-Haggin case began with conflicts over the lower end of the Kern River in the late 1870s. In the arid climate of the southern San Joaquin counties, especially after the drought of 1877–78, access to water was literally a matter of life or death for anyone trying to raise livestock or crops, and water rights were something to be fought over nearly to the point of war. In this vast, nearly flat semi-desert landscape, the major source of water was the Kern River flowing out of the Sierra into the sink of Buena Vista Lake and its associated Buena Vista Slough, a 50-mile-long marshy corridor angling northwest toward Tulare Lake. Like all California waterways in their natural state, lake levels and channel configurations in the slough varied from season to season and year to year. Landowners along the water courses, like Miller & Lux, staked their claims to water by riparian right—rights that Henry Miller had judiciously acquired in nearly all his land purchases. Other big landowners, however, like Haggin and Tevis, claimed water on the basis of prior appropriation—rights that had been granted in early California law to those who first began diverting water away from natural channels to use for beneficial purposes in mining and agriculture.

In the late 1860s Miller & Lux had arranged to buy some 40,000 acres along Buena Vista Slough, where the Kern River emptied out north of Buena Vista Lake, overflow land that qualified as swampland. By 1873 Miller & Lux controlled an extensive tract about 20 miles west of Bakersfield, marshland filled with tule grass and mosquitoes. Though the slough changed from year to year as weather conditions determined, it formed a fairly dependable oasis in what was otherwise nearly desert, providing a source of water for varied vegetation and wildlife. Titles to the swampland were not confirmed until 1876 and 1878, but to conform with federal stipulations about reclamation, Miller & Lux went to great expense to build a 21-mile canal to drain the slough area with cooperating neighboring landowners. The firm continued to expand its acreage along the general course of the waterway and eventually pastured thousands of cattle and other livestock there.

In 1874 James Ben Ali Haggin, a financier and speculator, arrived in big Kern County just as tens of thousands of acres of railroad land were being opened up for sale. Together with his partner Lloyd Tevis and their agent and associate William B. Carr, he was able to amass a huge swath of land in the Buena Vista Lake area. They formed the Kern County Land Company in 1875 with the professed intention of developing the raw land for sale to family farmers who would come later.

Carr, an ambitious political operative, helped Haggin and Tevis acquire close to 100,000 acres in a nearly solid block of holdings between the Kern River and the Buena Vista and Kern lakes fifteen miles to the south. Some of this was railroad land, some was designated as swampland, and some was acquired after the 1877 Desert Land Act, said to have been designed largely for the partners' benefit. With Tevis and Carr's close ties to railroad interests, the partners filed for 100,000 more acres adjoining the new Southern Pacific line north of Bakersfield. These land grabs were enabled by many dummy entrymen who were paid a pittance for their signatures. Meanwhile, a few pioneering cattlemen and farmers moved into the area in the 1870s too and began building small canals to drain land and direct water to where it was needed.

The Haggin partners soon owned more than 300,000 acres in Kern County. They arranged for water supplies by buying up stock in the recently constructed Calloway Canal upstream along the Kern River, then expanding it and several other small water conveyance systems in the area. These efforts provided jobs for dozens of men who were glad to have work on construction projects, and Haggin avoided public resentment of his land consolidations by pushing a public relations campaign that promised the desert would soon bloom for the benefit of small farmers. Nevertheless, he rarely visited Kern County.

Unlike Henry Miller, a hardworking German immigrant with few pretensions, James Ben-Ali Haggin came from a genteel southern background and after his arrival in California in 1850 moved in high society with other financial speculators. With his brother-in-law Lloyd Tevis he invested extensively in a variety of ventures, including Wells Fargo and Company and a couple of mining companies favored by George Hearst. Known for his luxurious lifestyle, Haggin had a liking for racehorses and owned the state's largest horse-breeding farm near Sacramento. Like the "Big Four" railroad magnates, he built an opulent home on Nob Hill in San Francisco, which occupied a whole block and boasted an 86-foot tower that could be seen from across the city. With wealth to spare, Haggin and Tevis acquired large tracts of valley land with an eye toward future turnover.

Their associate William B. Carr, on the other hand, was a brash, coarse frontier personality, who built his career first in the construction of mining ditches and levees and then became a supplier of brick for buildings in San Francisco, where he also forged political connections. "Boss" Carr was vilified regularly in the San Francisco press but exercised great influence in the distribution of official patronage. One newspaper identified him as part of "a ring of mercenary bandits," while another called him "the Napoleon of lobbyists and intriguers." He served as land agent and general factotum for Haggin and Tevis.

The severe drought of 1877–78 put heavy pressure on regional water resources. Haggin and Tevis began diverting large amounts of water upriver on the Kern through a newly built diversion dam on the Calloway Canal. Dozens of court cases were filed as a result of the drought, with plaintiffs arguing that upstream diversions were harming downstream users. The court decisions in these cases were mixed, some siding with the appropriators, some with the riparians down river. Meanwhile, vandalism was rife along the canals, and at least one armed confrontation faced off angry water users against each other. Miller & Lux did not go to court until after the death of some 16,000 cattle, most of them theirs, in 1878 and 1879 along Buena Vista Slough and near Tulare Lake, 20 miles below the Calloway dam. In May 1879, with seven other riparian owners, Miller & Lux filed for a judicial injunction against the upstream diversions of the Kern County Land Company. More suits followed in 1880 against other appropriators and ditch companies.

These legal battles over the limited water of an essentially semi-desert area aroused intense public interest. Although 1880 turned out to be a wet year, the Lux vs. Haggin lawsuit was brought to trial before the Kern County Superior Court in April 1881. While there were several other suits pending, the prominence of the parties in this case assured that it would become a battle royal, engaging the energies of more than a dozen attorneys and the attention of hundreds of newsmen and bystanders. The Miller & Lux attorneys included Chief Counsel Hall McAllister, a well-known San Francisco litigator, and his assistant R. R. Houghton, while Haggin and Tevis had an equally aggressive defense team with three leading attorneys including Louis T. Haggin, the eldest son of James Ben Ali.

The Kern County court heard laborious testimonies for 49 days. Called to the stand were several respected water engineers as well as government surveyors and county officials. A large portion of the Kern citizenry was questioned, including old timers with long memories, half a dozen Miller & Lux employees, and dozens of neighboring settlers with general knowledge about climate, water flows, and historical appropriations through local ditches. Some local characters like Captain John Baker, a riparian owner with a taste for quoting Shakespeare, were highly eccentric, adding to the color of the trial. It was said afterwards that the transcribed testimonies made a stack of paper four feet high.

Rather than argue the issue of riparian rights, which were endorsed in early California law, the Haggin team spent most of the 1881 trial contending that Miller & Lux were not truly riparian users—and that even if they were, the partners had not acquired title to their property until after the diversion dam was built, thus allowing appropriators precedence under the principle "First in time, first in right." Witnesses were called to give evidence on the questions crucial to the riparian argument: Did the Miller & Lux property include a flowing waterway, or was it actually swampland without a clear water channel? Was the Buena Vista Slough a natural channel of the Kern River, or was it a sink where the water simply spread out? The slough had few banks, water flow varied by season, and sometimes there was no flow at all. If there was no moving stream, how could there be a riparian right? These questions were minutely examined in dozens of testimonies.

Following the testimonies and days of technical legal arguments, the courtroom debates began in June. Both sets of attorneys pulled out the stops. The plaintiffs emphasized property rights and the common law endorsing riparian privilege, while the defendants played on popular antagonisms by posing the "cattlemen" (Lux) against the "farmers" (Haggin). Despite the fact that Haggin was a land speculator then operating a livestock business much like that of Miller & Lux, defense attorneys portrayed him as a "community builder" preparing the land for division into family farms. Miller & Lux, they contended, were selfish monopolists. Local newspaper editors tended to side with the Kern County Land Company even though they expressed mixed feelings; most viewed the company hopefully as an agent of economic progress that would ultimately benefit settlers. Miller & Lux, on the other hand, were considered stubborn "cattle rancheros" who only wanted big pastures for their herds.

On November 3, 1881, the Kern County Court found in favor of Haggin that "no continuous or defined channel" existed in the slough area, and that therefore the appropriators had precedence over the so-called riparians. Judge Benjamin Brundage ruled that the Kern River did not in fact flow through the Buena Vista swamp, that the Buena Vista Slough itself was not part of a flowing stream, that riparian rights could not properly exist in swampland, and thus that Miller & Lux could not claim them. If the wetlands were to be construed as an actual waterway, in fact, then the real riparian owner would have to be the federal government that still owned the dry land on the edges of the marsh! In effect, the decision endorsed land development through impoundment and diversion of water sources and viewed economic growth as dependent on irrigated farming.

Thus the local court had spoken. Meanwhile, however, in Tulare County next door, Judge W.W. Cross had ruled in late 1880 in support of a riparian owner's claim that the sink of the Tule River constituted a natural waterway even though the river itself virtually disappeared 10 miles above his land. In other cases too Judge Cross had consistently defended riparian rights in his county. Clearly the controversy was not yet settled.

According to newspaper reports in May 1881, Henry Miller, suffering from poor health, went east on the train to New York to recover. Miller spent months far from California and missed the drama of the Lux-Haggin trial in Kern County. In October, however, the *Gilroy Advocate* reported his return to California as "the picture of health." Arriving with new energy only to discover that the court had decided in favor of Haggin, Miller was determined to defend his firm's rights and immediately pressed for an appeal. Emboldened by the Tulare County decisions, he persuaded Lux to take the case to the State Supreme Court. Meanwhile the Legislature, pressured by constituents, was trying in its cumbersome way to grapple with water code amendments and new bills designed to reform the state's confused legal framework.

Another year passed, but in 1883–84 the Lux v. Haggin court battle resumed. Attorneys for both sides polished their arguments, piling up new testimonies and eloquently invoking God and Nature as well as legislative intentions and legal technicalities. This time Miller & Lux prevailed. In October 1884 the California Supreme Court overturned the lower court decision, ruling narrowly (4-3) in favor of Miller & Lux and declaring that English common law incorporating the riparian doctrine should take precedence over the legislation allowing appropriation. The court also made two major findings: that swampland could in fact carry riparian rights, and that Miller & Lux did in fact legally own their property before the Calloway Canal began appropriations.

The Supreme Court judges were so narrowly and sharply divided in their 1884 decision, however, that the case was appealed and heard for a third time, even as the Legislature was struggling with rival bills and anti-riparian interests were calling meetings to shape public opinion. On April 27, 1886, Judge E. W. McKinstry, a highly respected senior judge, gave a thorough, conclusive explication of the riparian doctrine and methodically disposed of each argument made by the Haggin team. This time the majority opinion emphasized that because the English common law had been formally adopted by the first California Legislature, the riparian doctrine and the principle of property rights took precedence over later legislation allowing appropriation of water to support beneficial off-stream uses. In his more than 200-page opinion—one of the longest in California legal history—McKinstry methodically addressed all points raised in the arguments of both sides. Nevertheless, his convoluted decision drew widespread criticism.

Though the riparians had won a major victory, court challenges continued, and so did legislative attempts at reform of water laws. A big anti-riparian meeting took place in San Francisco in late 1886, and a new organization called the "State Irrigation Convention," allegedly steered by Carr, began encouraging the formation of local clubs to fight the riparian doctrine and lobby for reform legislation. Activists even made an abortive attempt to reorganize the state Supreme Court in hopes of overturning the verdict. Bargaining and bribery by both sides were rampant in the late summer of 1886, when a special legislative session was called to discuss water reform. The session adjourned with frustration and bitterness but without resolution. Journalists weighed in with opinion pieces. The popular *Overland Monthly* featured a lengthy critical analysis by an anti-riparian writer in an article that excoriated the Supreme Court's recent "most extraordinary decision" as "too monstrous for belief." [3]

Years later a valley historian would sum up his view of the battle: "After the *Lux. Vs Haggin* decision of the Supreme Court in 1886, the Civil Code, Section 1410, was amended to read: 'The right to the use of running water flowing in a river or stream, or down a canyon or ravine, may be acquired by appropriation.' This was a direct challenge to the doctrine of the Supreme Court. Riparian rights were judge-made; appropriation rights were legislature-made. Irrigation law in California proceeded to march in double-time in both directions at the same time. No wonder that dams were built and blown up, that irrigationists and cattlemen fought and won and lost all at the same time, and that even the lawyers became dizzy." [4]

In 1887, in response to public agitation, the Legislature passed the Wright Act to facilitate the formation of public irrigation districts. Though this landmark law was written to allay water struggles between private owners, it was not uniformly effective. Some early public irrigation districts were fairly successfully organized, but others foundered; property owners continued to make new demands on water courses throughout the valley; and water attorneys prospered. The always watchful Henry Miller kept a sharp eye out for new appropriators upstream from the company ranches, directing his employees to patrol levees and report on suspicious activities. By century's end, thickets of ever-denser litigation and an increasing flow of court decisions into the public policy arena added to the total volume of the state's water laws, to nobody's great satisfaction except perhaps the attorneys.

Recognizing the continuing volatility of the issue, Miller pragmatically made a private settlement with the Haggin interests in 1888, agreeing to divide the flow of the Kern in such a way that both might benefit. Miller and his neighbors would be guaranteed unlimited use of Kern River water between September and February, and during particularly low water years in the summer too. The rest of the year Haggin would receive two thirds of the river's flow and Miller & Lux one third. Both parties would contribute to the construction of the Buena Vista reservoir to store any unused water. Meanwhile, Miller & Lux continued to buy up more Kern County land (120,587 acres by 1890), while Haggin, Tevis and Carr increased their holdings to more than 375,000 acres. Late in 1890 the Haggin associates began to offer 45,000 acres for sale in small parcels in four irrigation colonies, but these small farms did not flourish. Carr sold out his interest in the company about that time, and Haggin left California for truly greener pastures in New York. Henry Miller remained in the valley, his eye as always on his ranches.

The 1880s brought great changes in California. Mining, still a significant part of the economy, had become an increasingly engineered industry as well as a high-stakes gamble for well-heeled capitalists. Legislation regulating hydraulic mining ended some of devastation along the rivers draining to the Sacramento Valley floor, but thousands of acres had already been ruined, especially in the Yuba River watershed. Farmers were still growing wheat, but they were making less money as international markets shifted. As local irrigation systems were developed in the Central Valley, new plantings included orchards and vineyards as well as field crops like alfalfa.

Henry Miller was instrumental in getting a railroad line extended through the west side of the San Joaquin Valley, its route determining the site of the "new" Los Banos founded in 1889. After 1890 Miller & Lux shipped their cattle by rail, and the days of trailing their herds over Pacheco Pass were largely over. The firm played a major role in the development of the three towns Los Banos, Firebaugh, and Dos Palos, establishing several company businesses including stores, slaughter yards, lumber yards, hotels, banks, and public utilities.

The casual eye could see that the Central Valley landscape was radically altered, both in vista and in composition. Many of the natural curves and sweeps of land and water were nudged into the geometric patterns of human settlement. By 1890 millions of acres of virgin land covered with native grasses or oak woodland were plowed, cultivated, and fenced into rectangles and squares. Meandering rivers, creeks, and sloughs were impounded in artificial ponds or redirected into straight canals in order to control floods or channel precious water supplies. Wagon roads and railroads crisscrossed the country, extending further with each passing year. Farm corrals and outbuildings punctuated the long views, and farmyards became holding areas for crowded collections of metal implements as well as livestock. Pioneer structures clustered around convenient watering or meeting places that would later become towns. Often these plain little buildings were hastily constructed boxes, with or without porches, though some builders made modest attempts at architectural decoration.

The native people and the wildlife vanished or retreated to the hills as the valleys grew more populated. Wild herds of elk and pronghorn were supplanted by domestic herds of cattle and sheep. Millions of birds along the Pacific Flyway began to disappear in the San Joaquin as they had in the Sacramento Valley. Ducks and geese still gathered thick along the river floodplains during the winter, but market hunters now used heavy-duty shotguns to kill dozens of birds with a single blast. Some slaughtered them with small cannon just for the feathers.

Some farms and ranches still offered wildlife refuge. Henry Miller issued orders for protecting the few remaining tule elk on company property, and the survival of that animal today is credited largely to his protection. He also leased hunting grounds to duck clubs, hoping to encourage responsible hunting as a means of avoiding trespassing and vandalism.[5]

New horticultural species also changed California's natural patterns of vegetation, with mixed results. Overgrazing on rangeland allowed new and less nutritious forage plants to take hold, but settlers also introduced many successful new agricultural crops and began to plant trees for shelter and shade in places where nature had thus far provided neither. In some areas where native oaks were cut for firewood or construction, landowners tried to replace them with other trees, and progressive nurserymen brought in horticultural imports destined to do well in California, including various kinds of pine and several varieties of palm. It became popular for nineteenth-century builders to plant palms in the front yards of new houses, and many of the palms lasted longer than the houses.

The wholesale planting of eucalyptus ballooned in the 1870s after some years of experimentation.[6] As early as the 1850s eucalyptus seed had been brought into San Francisco from Australia, where these native trees in their hundreds of varieties served many purposes, and in the 1860s an Oakland nurseryman began distributing seedlings. By far the most popular variety was the Tasmanian Blue Gum (*Eucalyptus globulus*), a tall, fast-growing, handsome but messy tree smelling strongly of menthol. Elwood Cooper, a Santa Barbara rancher who also served on the State Board of Forestry and the Commission on Horticulture, touted the advantages of eucalyptus for shade, shelter, windbreaks, woodworking, and firewood, while the influential *Pacific Rural Press* promoted eucalyptus in numerous articles. Believers claimed that the trees could improve human health by counteracting bad air and draining mosquito-breeding marshy soils with their aggressive roots.

Many eucalyptus were planted along the coast, where their distinctive shape and height changed the entire look of formerly treeless slopes. But they were also popular in urban areas, providing shade and beauty with relatively little effort or expense. The university campus at Berkeley planted a eucalyptus grove that became a beloved setting for student ceremonial events. The Southern Pacific Railroad planted nearly a million seedlings along its California tracks. Nurseries offered seedlings for as low as 10 cents each, and organizations distributed free seed. Landowners planted blue gums for firewood or to line rural roads. Farmers and ranchers, including Miller & Lux, set out eucalyptus around their buildings and corrals.

The eucalyptus crusade faltered as critics grew disenchanted with the blue gums' litter and greedy roots as well as their tendency to die back in hard frosts. By the turn of the century the immigrant tree of such promise began to gain a reputation as an invasive nuisance. Between 1907 and 1913, however, enthusiasm returned during a new eucalyptus boom encouraged by forestry officials and university specialists, who believed the wood might augment declining hardwood supplies. Investors planted millions of seedlings in hundreds of groves intended to produce saleable lumber. This hope faded as the years passed, for the grain of the wood produced in California never equaled that produced in Australia, and problems with plantings appeared. In 1923 an official U.S. forestry report stated flatly that the eucalyptus craze was based on delusion. Worse yet, dense stands of the oily trees turned out to be a fire hazard. Eucalyptus litter in the Berkeley and Oakland hills contributed to destructive fires there in the 1920s and later.

Though some of the early hopes for eucalyptus were dashed, their legacy has lingered on. Eucalyptus trees still define many roadways and windbreaks, parks and campuses in California. It is hard to imagine the state today without this distinctive non-native that has so thoroughly altered the state's scenery—the icon of an era.

Away from the parry and thrust of the courtrooms, life on the Miller & Lux ranches had a steady predictability, ruled by the daily demands of animal husbandry, the changing tasks of the seasons, and by Henry Miller's own detailed directions. By 1880 the partnership's more than half-million acres in the San Joaquin Valley were divided into four main divisions, with a superintendent and crew for each: the Santa Rita division, also known as the Home Ranch, bordering some fifty miles of the San Joaquin River; the Los Banos Division to the west; the Posa Division around the small commu-

The Santa Rita Ranch near Los Banos was the primary headquarters of the Miller & Lux cattle company in the San Joaquin Valley. The cowboys were numerous, the haystacks huge.

Courtesy of Milliken Museum, Los Banos.

nity of Firebaugh to the south; and the Colony Division around the settlement at Dos Palos. These divisions were the core of the Miller & Lux empire, but the firm also owned dozens of smaller properties scattered through the coast range, in the Southern Division around Buena Vista Slough, and, as the decade wore on, in Nevada and Oregon.

While innumerable head of cattle, horses, sheep, and hogs grazed the pastures, the company payroll might include anywhere between 300 and a thousand employees, according to the season and the labor needed. Each division provided room and board for seasonal employees. The company built big bunkhouses to house up to 40 or 50 working men at a time. Consisting of large open rooms filled with beds, the early bunkhouses had a stove in the middle surrounded by boxes for seating, and at first the men furnished their own mattresses and bedding. Later Miller directed that they should be given sheets, and "when possible have a room set apart where the men can talk or read at night. Such a room needs a lamp and a stove." Permanent employees with families were assigned to individual small houses, while most of the year-round single men had quarters at the Canal Farm.

Ranch workers were usually divided into groups that specialized in various activities. Many worked directly with the livestock, but others were engaged in reclamation or farming–digging ditches, building levees, throwing up irrigation checks, directing water, clearing roads, running and repairing fences, planting and harvesting feed crops–all of the varied tasks necessary to the development and upkeep of big ranches. Common laborers generally made $1 a day, sometimes less. The division of labor was often based along ethnic lines for convenience' sake, for as new immigrants came into the valley seeking work on farms and ranches, they began appearing on Miller & Lux payrolls, and work crews did better if they spoke a common language.[7]

Throughout the early years of the firm Mexican vaqueros dominated the cattle operations. Men who had learned herding, roping, and branding from their families' long experience in California often knew the landscape intimately as well, and their skills at separating and moving cattle were invaluable. They knew the places where cattle could ford waterways and where stock had a tendency to migrate; they knew how to handle frightened animals and how to nudge herds into temporary or permanent enclosures. These Mexican vaqueros had a special status on the ranches. They were steady employees and relatively well paid, and some worked for Miller & Lux for many years. Most spoke only Spanish among themselves, though the top hands always spoke English too, and many were related by blood–fathers, sons, brothers, uncles, cousins–which helped make the crews cohesive. Two of the firm's most highly respected vaqueros were Tomas Diaz, who worked for Miller for more than 20 years beginning in the late 1850s, and Rafael Cuen, who served as head vaquero for the Southern Division for 35 years. As stock farming and railroad shipping replaced the old cattle drives over open land, however, these highly skilled vaqueros grew less needed. By the 1890s some of the older Mexican vaqueros were succeeded by younger Anglo "buckaroos," marking an end to the former largely Hispanic cowboy culture.

As in other parts of the state, Chinese workers became available in the San Joaquin Valley after the completion of the Central Pacific Railroad. Already conveniently organized into large work crews supervised by their own bosses, they provided contract labor during the building of the San Joaquin and Kern River Irrigation Canal in the 1870s, at one point outnumbering white workers by three to one. As anti-Chinese prejudice grew and Chinese crews simultaneously began pressing for higher wages, however, Henry Miller became apprehensive about their potential ability to strike and directed his supervisors to look for other sources of labor.

Large Chinese crews continued to provide manpower for other big projects and were hired to reclaim marshland and build levees in the Delta, but some individual Chinese began to lease small plots of land for their own intensive farming. Many Chinese moved into domestic work, where they made reputations as excellent cooks, "houseboys," and laundrymen. As in other rural areas where women were in short supply, the Miller & Lux ranches hired dozens of these "Celestials," as they were nicknamed in the press. In 1888 Henry Miller remarked to an interviewer, "I suppose we have 50 cooks, and 49 of them are Chinese." And he added, matter of factly, "The Chinese are a necessary evil, if all of our people would work there would be no Chinamen among us."

Like Miller himself, the Chinese were habitually frugal, trained through generations to make the most of every resource, and as a rule they were excellent workers, putting in long hours without complaint and making few demands. Often ingenious and always industrious, they adapted their cooking to western tastes while closely managing their kitchens. In many affluent city homes the Chinese were valued servants for years, sometimes considered practically members of the family. On the big Miller & Lux ranches they came and went, living and working in isolated circumstances but acknowledged as nearly indispensable until they voted with their feet and moved on.

At about the time that anti-Chinese legislation in the 1880s began to take effect, cutting the number of immigrants from Asia, newcomers from Europe came in growing numbers to California. French immigrants tended to move into San Francisco, while Germans continued to spread out into the farming districts. Italians planted vineyards or grew vegetables on truck farms, and Portuguese started small dairies in the Bay Area, the San Joaquin Valley, and the North Coast. Contingents of eastern Europeans, including Croatians and

Serbs, also appeared in the San Joaquin during the 1880s. Many arrivals hoped to settle on small farms, and some participated in colony settlements. Armenians fleeing political persecution came to Fresno County to begin raising fruit that they knew from their homeland, contributing raisins, figs, and melons to California's mix of crops. But many immigrants without financial resources went to work for big employers like Miller & Lux at least temporarily. Italian names appeared on the firm's payrolls with regularity. Most worked as seasonal laborers, usually as irrigators or hay harvesters, until they could find better work. By the 1890s more than half the seasonal workers at the Dos Palos Division were Italian. Other European immigrants appeared frequently on company lists too.

Miller & Lux operated four divisional dining rooms serving hundreds of employees. Henry Miller insisted that all his workers should be well fed with a variety of plentiful and wholesome food. The kitchens served lots of meat, including pork and mutton besides copious amounts of beef, and offered many kinds of fruits and vegetables produced in ranch gardens and orchards. "Good vegetables make a man more contented and draw a better class of laborer," wrote Miller to his ranch managers. While the superintendents and professional employees had special dining areas and got somewhat more elaborate food, everyone ate well at the Miller & Lux ranches, and visitors were always invited to join in for meals.

Over time the Miller & Lux ranches in the San Joaquin became known as the "dirty plate route" for hoboes moving through the valley. After a series of incidents of vandalism or arson on local ranches that refused to give food to tramps, Miller issued standing orders that men asking for food should be allowed to have a free meal in the company dining room, but told that they would have to eat after the employees on the plates left behind in order to save

The migrant workers or hobos looking for work on farms and ranches often traveled long distances carrying their own bedrolls and in popular jargon were called "bindlestiffs." *Courtesy of the Bancroft Library, University of California, Berkeley.*

on dishwashing. Tramps were happy to get good food and complied with this rule, and as a result the Miller & Lux ranches were rarely troubled with gates left open, carelessly or maliciously set fires, or other such problems. Poso Ranch records during one February in the 1890s reported that transients received 1,028 free meals, or about 36 a day. A figure of this magnitude suggests hard times, but it was known for years that hoboes planned their routes to coincide with mealtimes at Miller & Lux. Edward Treadwell wrote, "Often after a hot day you could see a tramp coming out from under a bridge, another from under a stop gate, another from a clump of willows, another from behind the barn, and all of them would gradually work their way toward the dining room door. Sometimes as many as eight or ten would be on hand at a single meal. Sometimes the foreman would have to stand by the door with a club to keep them out until the men were served. But generally they understood the rules, one meal, one night, and no more." [8]

The "dirty plate" policy was only one of Miller's many explicit directions on ranch management, which he gave in person during his frequent visits or wrote out in numerous letters addressed to superintendents. Shortly before Miller's death Paul Parker, a long-time employee of the firm, compiled a lengthy reference handbook for the use of Miller's successor, Leroy Nickel, and his first general superintendent, Ralph Merritt. Organized into sections and neatly typed, this remarkable document, nearly 200 pages long, contains a mass of observations and instructions found in hundreds of old letters stored in the basement of Miller's San Francisco home, written over many years to his managers in California, Oregon, and Nevada. As edited by Parker, the guidelines range from "Seasonal Warnings and Monthly Reminders' to specific directions on dozens of subjects ranging from the production of alfalfa to the care of young horses to methods for smoking bacon and ham. Distilled from years of observation and experience, they reveal Miller's range of knowledge as well as his insistence on micro-managing his far-flung domain.[9]

"Horse manure is very good for land that is spotted with alkali," he wrote, for example, while also recommending strychnine for squirrel problems, raisins for hog feed, the covering of haystacks against bad weather, ways to grub out cockleburs in fields, and hundreds of other daily and seasonal practices from the corrals to the kitchens. In addition to these always sensible instructions, Parker printed Miller's "Laconic observations," including such pragmatic pearls as "It is no disgrace to ask advice from one holding a position below you" and "A man slovenly in his surroundings is the same way in his work." Miller's basic philosophy is revealed in his oft-reiterated maxim, "Fortunes are made by taking care of the small things."

At age 60 Henry Miller was very well known and widely respected for his business acumen. He had been homely as young man, with a drooping mouth and receding hairline, though his eyes were bright and piercing. He grew better looking as he grew older, growing a short, neatly trimmed beard and generally wearing dark suits of an old-fashioned cut with a black bow tie. He was scrupulously clean in his clothing and habits, took no drink or tobacco in an age when both were widely abused, and ate simply. He was "strong as an ox," with a stocky build and jutting chin. Even at 60 he was tough as nails, though by this time he had given up horseback riding to visit his ranches by buggy.

His passion was work. His personal life was quiet, and he lived relatively modestly. He indulged in few luxuries, had no taste for elegant entertainments, and had few friends. Because the challenges of business continued to engross him, he took little time for anything else, even his family. He rose from bed before daybreak and traveled thousands of miles each year in the heat and dust or the rain and cold.

Miller was proud of being a self-made man, and he could be very demanding. He was famous for a fiery temper when thwarted, and many local storytellers recounted anecdotes about his throwing his hat to the ground and stamping on it when he was particularly infuriated. Curt and incisive, he tended to be humorless and autocratic in his transactions, yet despite his impatience he was usually courteous in his business dealings and both fair and considerate toward employees who pleased him. When he thought someone worthy, he was known to be generous; when he thought it strategic, he was known to give bribes.

It was a sign of his fame that he was interviewed for a biographical sketch by the History Company of San Francisco, which led the trend toward publishing "mug books" of prominent Californians, bringing out Bancroft's *Chronicles of the Builders of the Commonwealth* in 1891. Interviewer Edwin Fowler took down Miller's dictation, which includes many of his thoughts on achieving success and reveals both his determination and his own strong ego. In his transmittal letter to the publisher Fowler notes that Miller is a "rusher"–a rapid talker, assertive and concise–a man of "indomitable will" and "thoroughly imbued with the spirit of progress," having great powers of concentration, "a fabulous memory, brilliant organizing power, and a fanatic devotion to work." Fowler also describes him as moral in outlook and abstemious in habits, but without religious belief.[10]

The loss of his youngest child, Sarah Alice, had been a great blow to him, yet Miller's relationship with his other children was problematic. His son Henry Miller, Jr., did not inherit the strong constitution of his father, and the father, traveling constantly on business, paid relatively little attention to him. The boy grew up largely in San Francisco and influenced by the affluent youth of the city. At an early age Henry, Jr., became a heavy drinker, and he was rumored to have contracted syphilis at a time when venereal disease was widespread. In 1889 Miller learned that his son had secretly married a female friend, Maude Oldham, and became so angry that he publicly disowned him in a newspaper announcement (he later retracted the decision). The son occasionally visited the Miller & Lux ranches but seems to have had no role in the business. After his first wife died young, he married again but never had children.

Miller's surviving daughter Nellie married James Leroy Nickel in a society wedding in 1885. She was 20, he was 14 years older. Nellie became a lady of fashion, occupied with her home and family in San Francisco and taking relatively little interest in the business that provided life's luxuries. Miller is said to have disliked his son-in-law ("When he comes in the room, I go out."), but the couple presented him with four grandchildren, and later, as Miller grew older and unable to continue his iron grip on the firm, he transferred authority to Nickel as the president of Miller & Lux.

In 1888 Miller expanded the living quarters at Bloomfield Farm by building a 44-room, three-story ranch house designed to house several employees as well as members and guests of the Miller family. This sprawling structure, as big as a hotel, featured 19 bedrooms, several long verandahs, a tower, and a distinctive colorful striped roof. For years the extended Miller family and their guests enjoyed spending time there. Miller also developed a summer retreat at Mt. Madonna, a cool and beautiful spot in the coastal redwoods in the mountains west of Gilroy, where the air was fresh and the heat of summer left below. The Miller family began summer camping there in 1879, but it was never a wilderness experience. Colorful tents were furnished and carpeted, then later converted to cabins for family and visitors. Mrs. Miller was known as a liberal hostess, and the compound eventually included a dance hall for entertainment. On Sundays the family often entertained 30 or 40 guests at afternoon barbecues. In his later years Miller built a seven-bedroom house on the property, featuring craftsman stone work and broad porches, and took great interest in the landscaping.

Charles Lux died on March 16, 1887, of pneumonia after a period of ill heath. He was 64 years old. His obituary, including a large pen-and-ink sketch derived from a portrait, appeared on the front page of the *San Francisco Call* the following Sunday. Called "a noted citizen of San Francisco" with "an enormous landed estate accumulated by a life of earnest toil," he was also described as "a thorough German, patient, persevering and determined." The *Call* traced his

Henry Miller's Bloomfield Farm near Gilroy served as a holding area for cattle coming from the San Joaquin Valley to city markets. The big house with its colorful roof was built in 1888.

Courtesy of Gilroy Museum.

career from his arrival in San Francisco as a butcher's assistant through his fabulously successful partnership with Henry Miller and recalled him as a generous and kindly man who had recently personally delivered Christmas turkeys to several dozen friends in the city. It was reported that Lux's will left more than $100,000 in bequests. Half his property was left to his widow, the other half in varying shares to his brothers, sisters, nephews, and nieces in Germany. His executors were named as Henry Miller, Miranda Lux, and his stepson Jesse Potter.

Henry Miller was now the sole surviving partner of Miller & Lux. More than a decade before, in June 1875, Lux and Miller had signed an agreement, apparently at Lux's instigation, stipulating that after the death of one partner the other would be allowed to manage the business for up to seven more years before final liquidation of assets. This agreement was superseded in 1884 by another more detailed version giving total control over all the real and personal property of the corporation to the surviving partner. Heirs of the deceased would receive their fair share of the profits not needed by the business, but the surviving partner would be the sole judge of those needs. Although the partners were in good health at the time of the agreement (Lux had just turned 60, and Miller 57), the wording of the revised contract suggests that it may have been Miller who wanted to make sure he would still be in command–legally as well as functionally–if his partner died.

Thus, in 1887 after Lux's death, the partnership became a quasi-corporation, with half the partnership going to each side. In the reorganization process Miller took complete charge of all operations, including the offices in San Francisco as well as the outlying ranches and subsidiary properties. He recused himself as the executor of his partner's estate, leaving that task to Lux's widow and stepson, but for the next seven years Miller kept an iron hand on the helm of the firm. Not surprisingly, he opted to plow most of the reported profits back into the business, despite protestations by the Lux heirs, who were hamstrung by the stipulations in the 1884 agreement. Later Miller would become the sole owner of the corporation, but not before a great deal of pleading and sniping in a series of court contests lasting 20 years.

In their 30 years of partnership Lux's geniality and diplomacy had made him the associate and friend of the business elite of San Francisco, enabling him to tap financial resources that bankrolled Miller's years of prodigious land buying and ranch improvements. Though Miller was always the driving force behind the partnership, he could never have succeeded in his ambitions without Lux to support his efforts in their early years. Politicians, bureaucrats, and journalists alike had admired Lux's amiability and tact, characteristics not noted in Miller.

Now, however, despite Lux's death, the firm was about to enter a new period of expansion across state lines, extending its reach into Oregon and Nevada. Miller alone would be responsible for that success.

1 M. Catherine Miller, *Flooding the Courtrooms,* reviews much of the litigation over the years.

2 Accounts of the Lux vs. Haggin case are given in Pisani, *Water, Land, and Law in the West;* Igler, *Industrial Cowboys;* and Bremer, "The Trial of the Century...Lux v. Haggin," *Southern California Quarterly.*

3 Marks, "The Riparian Decision in Interior California," *Overland Monthly,* 1887.

4 Smith, *Garden of the Sun*, p. 637.

5 The Frank Latta collection at the Huntington Library includes many items relating to San Joaquin history over decades.

6 See Santos, *The Eucalyptus of California*. Hilgard included a section on eucalyptus in his 1882 College of Agriculture *Report*, pp. 18–22.

7 Igler, *Industrial Cowboys*, pp. 124–44, describes Miller & Lux employment practices in line with his argument on industrialization.

8 Treadwell, *The Cattle King,* pp. 324.

9 Copies of the handbook of Miller's instructions are preserved in the California Historical Society Library and the Huntington Library.

10 Fowler interview reprinted in *One Man Show.*

~9~

The Oregon Connection

THE NORTHWESTERN PORTION OF NEVADA and the southeastern portion of Oregon together are composed of an immense, sprawling volcanic plateau nearly a mile above sea level, part of what is known as the Great Basin country. This is rugged and austere country, broken by occasional mountains, mainly covered with sagebrush, sometimes studded with juniper, generally arid but producing very nutritious bunch grass in areas of sufficient water. The contrast between the predominant rolling gray-toned sagebrush hills and the occasional small green valleys is striking. The land has limited resources and hard winters, but from the 1870s on the region gained a reputation as fine cattle country. Western stockmen began to appreciate the fact that cattle raised in the cooler climate of the high inland plateau were healthier and less subject to disease than those raised in the hot valleys of California, and that the wide open spaces allowed the continuation of a traditional livestock culture well after it had begun to fade in other areas.

Even before 1880 Henry Miller, concerned about the growing incidence of Texas fever in the San Joaquin, began to cast his eye on the possibilities for expanding his network of ranches to include units to the north. He had already tried buying cattle from the interior, but he had been less than pleased with their condition on arrival in California and was naturally inclined to exert more control over further imports. Rangeland in remote Nevada and Eastern Oregon was still very cheap. Miller began to envision a chain of strategically linked properties that might allow him to develop his own carefully managed cattle production on the high plateau. As always, he waited until circumstances might coalesce in his favor.

Miller's first land purchase outside California seems to have been the Nevada ranch offered by the Hugh Glenn estate to help settle debts after Glenn's death in 1883. This was a spread in Washoe County west of the forbidding Black Rock Desert country, a land of dusty playas, the dried-up beds of ancient lakes. Between two mountain ranges some fifty miles north of what is now the little town called Gerlach–which did not exist until the railroad came through the area in 1908–the springs of a surprising oasis called Mud Meadows had provided much-needed water and grass for pioneer trains along the difficult Applegate Trail into Oregon. The place also served for a few years as headquarters for an army post, afterwards becoming known locally as Soldier Meadows. After

The long chain of Henry Miller's cattle ranches in the San Joaquin Valley eventually grew to include many properties in southeastern Oregon and northwestern Nevada. This map—showing the widely scattered landholdings—appeared in the 1950 edition of Edward Treadwell's *The Cattle King*.

OREGON
SILVIES VALLEY
Silvies River
AGENCY VALLEY
Malheur River
VALE
HARNEY VALLEY
IRON MT.
STEENS MT.
ALVORD DESERT
LAKEVIEW
IDAHO
FT. McDERMITT
Quin River
BLACK ROCK DESERT
WINNEMUCCA
CALIFORNIA
Sacramento River
NEVADA
PISEN SWITCH
ALKALI LAKE RESERVOIR
POWER PLANTS
N FRANCISCO
San Joaquin River
MT. MADONNA
GILROY
PACHECO PEAK
San Benito River
GREAT CANAL
Kings River
Tulare Lake
Kern River
CARRISA
Buena Vista Reservoir
LOS ANGELES
PACIFIC OC

the state of Nevada began selling land in that area in the 1860s, Hugh Glenn had purchased acreage for raising livestock there with his partner E.W. Crutcher, but its isolation and limited assets had prevented the remote ranch from becoming profitable. Doubtless Miller was able to buy it at a low price. Miller's management began to improve it, and he later bought up other properties in the area, combining them under the general name of the Black Rock Ranch.

Miller might well have hoped to acquire Glenn's great P Ranch in southeastern Oregon too—another remote property in Harney County's Blitzen Valley, about 100 miles north of the Nevada border—but that was a flourishing place being competently run by Peter French, who for the time was also married to Glenn's daughter, and it was not for sale.

In the mid-1880s, however, Miller began collaborating with another rancher in the Oregon high country, Tom Overfelt, whom he had met at a cattle sale in San Francisco. He was sufficiently impressed with Overfelt to encourage him to raise cattle for shipment to be finished in California by Miller & Lux. With Miller's offer of financial backing, Overfelt & Company (Miller being the silent partner) began to gather up scattered homestead properties in Harney and Grant counties, in the Silvies and Malheur river watersheds—small ranches in the Silvies Valley, along Trout Creek, near Drewsey and Warm Springs, and at Kimball Flat, Harper, Pine Creek, Stinking Water, and other places strategically positioned along the tributaries that reach like fingers down from the Blue Mountains. Overfelt was also able to acquire some choice land in the Agency Valley, a 2000-acre valley along the north fork of the Malheur River east of the tiny settlement of Burns, after that valley was opened for land claims following the closing of an Indian reservation there. Each of Overfelt's little ranches stationed along the waterways had grazing rights in the surrounding public domain and thus had access to perhaps ten times as much land as was owned outright, while the Agency Valley already had a rudimentary canal system for irrigating pastures and feed crops. After Henry Miller visited the area he could see how the Overfelt holdings might fit into the growing Miller & Lux cattle empire, even though they were well over 500 miles from San Francisco.

Overfelt was an energetic and competent cattleman who got on well with Miller, but he died unexpectedly in a horseback riding accident in 1886. His company was thus dissolved, and Miller & Lux bought up his assets and his "LF" brand (with the F positioned backwards), reincorporating the business as the Pacific Live Stock Company, known locally as the PLS, in 1887. Anticipating weather problems on the high plateau, which was known to be a region of harsh winters, Miller directed his PLS ranch managers to grow and stockpile quantities of hay and build sturdy winter feedlots for their herds. Consequently most of the PLS cattle made it through the following three hard years of drought and cold.

Henry Miller was waiting in the wings when another southeastern Oregon cattleman, John Devine, got into financial trouble. During the 1870s Devine and his partner W. D. Todhunter of Sacramento had acquired some 150,000 acres in the sagebrush counties. Their first property was the abandoned site of Camp C.F. Smith, a former army outpost in a small oasis south of Steens Mountain, where four creeks flow down from the heights to water a little valley. There Devine put up several stone buildings and a large red barn with a cupola surmounted by a weather vane replica of a white horse. This place became known as the White Horse Ranch. Devine and Todhunter then teamed up with a land speculator, Henry C. Owens of Eugene, who manipulated the laxly administered land laws of the time to claim Harney Basin "swampland." Their land acquisitions included the lush meadows of the later-named Island Ranch, which lies between two forks of the Silvies River south of Burns, plus the fertile "Red S" fields at the lower end of the Silvies just north of Wrights Point. State land maps had indeed designated these areas as swamp or overflow lands in periods of wet weather, although for most of the year the climate and the high-desert soils are dry. Todhunter & Devine also acquired small properties at Juniper Lake, Mann Lake, Alvord, Sarrino Point, and elsewhere east of Steens Mountain.

Devine was a colorful, vain, and arrogant character much addicted to fine horses and expensive Spanish-inspired riding costumes. He antagonized many with his high-handed ways. Although he was able to accumulate a great deal of land, he was not a particularly good cattleman, nor a businessman. Todhunter & Devine found their ambitions foiled by the series of dry years from 1887 through 1889, which were followed by the killing winter of 1889–90, a period that dramatically changed livestock ranching on the high plateau. Cattle weakened by hunger died by the thousands in the bitter weather, bringing many ranchers to the verge of bankruptcy. John Devine's cavalier style of management and his financial setbacks from several land litigations combined with the extreme weather to force him to sell out.

Thus in 1890 Miller's Pacific Live Stock Company absorbed the Todhunter & Devine properties and the company's "S Wrench" brand for $1 million. Miller, feeling some sympathy for the rancher's plight, offered Devine the opportunity to stay on to manage his former holdings, but Devine had little aptitude for being the employee of another, and the relationship soon soured. Miller later remarked that he had never met a man who could make so many enemies, but he nonetheless gratuitously deeded back to him the 6,000-acre Alvord Ranch lying below the highest point of Steens

Mountain. There, looking out over the Alvord Desert., Devine lived until his death in 1901.[1]

Drought and severe winters brought down another big rancher in western Nevada. Miller was ready to take advantage of the opening there too. Miller & Lux acquired the ranches and the "7 S" brand of N. H. A. "Hock" Mason, who had accumulated many thousands of acres in a domain that extended from the Walker River east of Lake Tahoe to the Oregon border. Miller had earlier encouraged Mason to raise cattle to ship to California, in effect assuring him a market, and promised to help finance him in establishing ranches. Mason had subsequently bought out a number of homesteaders, first along the Walker River in a valley now called Mason Valley, south of Reno, in the "Pizen Switch" area where two trails split off toward California and western Oregon, not far from railroad shipping facilities. Mason then extended himself further north to acquire widely scattered lands in the Quinn River country south of the Oregon line and in the country bordering the Black Rock desert. Some of these properties were paid for by the famed "Bullhead" drafts on Miller & Lux, then recognized almost everywhere in the West as valid currency nearly equivalent to U.S. cash. These small ranches scattered across northwestern Nevada were surrounded by hundreds of thousands of acres of the public domain, where cattlemen could still run their herds across free rangeland.

In this isolated open country, cattle were usually put out to range in the spring and summer but brought back into the ranches during the winter to feed on hay. Mason did well at raising cattle on his holdings at first, but before long he unwisely allowed his herds to expand beyond the haying capacity of his ranches, relying on the surrounding rangeland to keep them healthy. During the dry years of 1887–89 his ranches put up very little hay. Then came the killing winter during which thousands of cattle died, and Mason went broke. The Pacific Live Stock Company absorbed the Mason assets at a fraction of their worth. But then, respecting Mason's abilities, Miller hired him to continue on as manager.

Henry Miller himself spent considerable time in Nevada visiting the ranches extensively to assess their management and reorganize them under his own more effective guidelines. The rundown properties were gradually revitalized. Knowing that the key to success was control over water, he arranged to have all the original property owners testify about their water rights, and these depositions were duly placed in courthouse files. He also sent agents out to locate all the unclaimed springs in that dry country and afterwards bought title to them with land scrip. Thus Miller & Lux eventually acquired title to about 90,000 acres in Nevada, all with documented water rights, plus access to public land over thousands of square miles. Ironically, whenever there was litigation over water rights involving Miller & Lux in Nevada (and there frequently was), the firm usually argued for those rights based on the principle of prior appropriation.[2]

The Pacific Live Stock Company thus became a leading player in the far western cattle industry, operating in southeastern Oregon and northwestern Nevada for some forty years as the inland arm of the Miller & Lux empire in California. Henry Miller had moved beyond the great Central Valley onto the high plateau of the Great Basin country, putting his distinctive stamp on ranch management there even as the world changed around him.

During the 1890s the Sacramento Valley holdings of the Glenn family grew less profitable as the wheat bonanza faded, but in Oregon's Harney County the French-Glenn ranch was still being

This is the only known portrait of Peter French, the manager of the P Ranch in Harney County, Oregon.

Courtesy of Oregon Historical Society.

capably run by Peter French and producing enough quality cattle to provide a steady stream of income for the owners in California. Despite his divorce from Ella Glenn in 1891, French remained remarkably loyal to the Glenn family. Even after the incorporation of the French-Glenn Livestock Company in 1894 diminished his own financial stake in the P Ranch enterprise through a lopsided division of shares, French did his best to manage day-to-day operations. He lived quietly at the ranch except for the seasonal drives of cattle to market and annual visits to California. His account books recorded continuing transfers of money from Oregon into California.[3]

But times were changing here too. An influx of settlers between 1880 and 1890, reflecting the national movement westward, increased the tiny population of big Harney County nearly 600 percent, from a scattered 419 persons to 2,559, most of them living in or near the few communities like Burns or Drewsey (formerly known as Gouge Eye, a name rejected by the postal authorities). New settlers trying to make a go of it on these austere high plateau lands sometimes found the life too challenging, and many gave up after a few years of struggle. The large stockmen who had come in early already controlled the most favorable lands and watering places, limited as they were; and these cattle barons became more aggressive in their own land acquisitions as they tried to deflect incoming homesteaders, taking advantage of the state's land law loopholes.

Like other big cattlemen, Peter French himself fenced in some 30,000 acres of the public domain, some of it illegally, and in trying to consolidate his P Ranch holdings during a ten-year period bought up thousands of new acres. By the mid-1890s the French-Glenn properties totaled 105,000 acres on the tax roles, most of it in the Malheur Lake area, while Henry Miller's Pacific Live Stock Company claimed about 83,000 acres to the north and south. These big companies also controlled two to three times as much public land through strategic fencing and positioning of their herds. Prospective settlers coming into the region found miles of sagebrush on rolling hills overarched by a splendid sky, but they were frustrated in locating fertile land near reliable water for homesteads. The big cattlemen seemed to have it all.

During the early 1890s some of the newcomers' resentment broke out in a rash of hit-and-run haystack fires and nocturnal fence cutting on the big owners' ranches. By the middle of the decade, hostility came clearly into the open, following several years of controversy about the status of land in the shoreline area around big, shallow Malheur Lake. A basin lake with no outlet, formed largely by drainage patterns from the uplands, the upper and lower waterlines of Malheur Lake varied naturally in level with the seasons and with the years. Human activities, including French's own irrigation efforts, had combined with a series of dry years to drop the lake line further than ever before seen. Would-be settlers believed—or hoped—that the land thus uncovered should be defined as new land open to settlement. Although the French-Glenn Livestock Company owned the surrounding area, a few squatters hoping to preempt this "new" land moved in on some of the thousands of acres exposed between the former water's edge forming the boundary of the P Ranch and the latest shoreline formed by the receding water. Here they hastily threw up little buildings and corrals and brought in small numbers of livestock to try to validate their claims.

In earlier years Peter French had done his best to get along with neighboring small landholders by offering them occasional work at the P Ranch, giving them the opportunity to supplement their incomes while learning something about his own ranch management. His hired hands, permanent or temporary, generally liked and respected him, as did the townspeople in Burns with whom he did business. Squatter animosity was rising against the big cattlemen, however, and as French continued trying to consolidate his own hold over land, feelings began to change. People in Burns, even those who admired French, tended to sympathize with the small ranchers who were struggling to make ends meet, and ill feeling was doubtless fanned by discussions in the several saloons where hotheads lingered. In the context of the times—the populism of the 1890s was a powerful political force even in remote southeastern Oregon—it is not surprising that the bitterness over the land question erupted in vandalism and occasional violence.

In August 1894, after a meeting with the French-Glenn company stockholders, French sent out letters formally requesting the squatters to vacate the acreage they were occupying in the lake meander zone. When the squatters did not respond, French filed several lawsuits in the federal court at Portland. Hoping to save time and expense on both sides, French suggested that just one test case should be forwarded to represent all the defendants, but two years passed before the French-Glenn vs. Springer case was finally brought to court. The case was heard in the Burns circuit court in May 1897. The key question in the deliberations was to determine who had rights under the law to the exposed land. "... did the receding water of Malheur Lake, by riparian right, extend the boundaries of the plaintiff's lands—or open a new region for settlers? Common law would have brought in a verdict for French-Glenn. The jury in the circuit court at Burns decided in favor of the defendant, Alva Springer," wrote author Giles French (no relation to Peter) in his narrative of the trial more than half a century later.[4]

Despite this apparent victory for the squatters, their antagonism only increased when they learned that French planned to appeal the case to a higher court. Unlike French, few of them had the financial means to keep pursuing the matter. Worn and bitter, they murmured among themselves. It was only a matter of months before ill feeling broke out into violence.

The day after Christmas, 1897, in a case eerily reminiscent of that of Hugh Glenn nearly 15 years before, Peter French was murdered in broad daylight by an angry man with a grudge. French had just returned from a business trip delivering cattle by rail from Winnemucca to Chicago. Back on the P Ranch, he was out herding cattle with his men when a local settler named Ed Olivier appeared unexpectedly from a swale riding rapidly toward him on horseback. The two men had some history. Although Olivier had not been involved in the suits over the Malheur Lake acreage, French had tried to deny him an easement through P Ranch land for a road to his homestead. Olivier had eventually received the easement from the county anyway, but friction between the two men had grown, and on at least one occasion the two had been heard exchanging angry words.

According to witness accounts by P Ranch employees in the later court trial, Peter French had just remounted his horse after opening a gate when Olivier came charging toward him, spurring his mount directly into French's and causing the animal to stagger. French had only a small homemade willow whip that he had been using with the cattle, but he tried to defend himself with it, first hitting at Olivier's horse and then at Olivier himself when the man tried to charge him again. Olivier then pulled out a gun, and French reined his horse away. As he turned to look back at his attacker, Olivier shot him in the head from a few yards away, hitting him in the right cheek. The bullet tore through French's skull, killing him instantly, and he fell to the ground. Olivier wheeled his horse around and rode away at a gallop while the stunned men of the P Ranch crew gathered around their leader's body in the field.

Within a matter of hours Edward Lee Olivier was arrested and imprisoned. Funds for his defense were raised in the community, and his trial for murder took place in Burns in May 1898. Though the evidence clearly suggested that Olivier, a man with a previous record of violence, had assaulted French without warning, the defendant made a plea of self-defense. His defense attorney pulled out all the stops, appealing along the way to general sympathy for careworn homesteaders and their families (placing Olivier's wife and children prominently in the front of the courtroom) and playing on popular resentment against the big cattle barons. In what seems to have been a miscarriage of justice, the Burns jury let the murderer go free.

The trial marked the tragic end of the career of a major figure in Eastern Oregon history. Peter French's body was taken to California to be buried in his family's plot at Red Bluff. As for Ed Olivier, his swagger made him disliked in the community, and his wife divorced him within the year. He left the county with another woman and what remained of the money that had been raised in his defense, never to be heard of again. The murder itself became part of local folklore, recounted in later years from varying points of view.

The great P Ranch itself never recovered from the loss of Peter French. With its management now uncertain, the ever-aware Henry Miller might have liked to acquire the property for his Pacific Live Stock Company, but F. C. Lusk, the administrator of the Glenn holdings, arranged to have a well-known neighboring rancher, Bill Hanley, oversee the ranch for several years. The P Ranch suffered under Hanley's hand, however, and debts began to accumulate. After some bouts of litigation the Glenns decided to sell the land in 1906 to the investment company of Henry Ladd Corbett, a prominent banker in Portland, and his partners. By then Henry Miller was on the verge of retirement.

In Oregon and Nevada, the outer reaches of the Miller & Lux empire, the Pacific Live Stock Company operated inland on the high plateau as a semi-autonomous subsidiary of the main company. Along with its other holdings along the scattered waterways of Harney County, the company came to own the greater part of the greener Silvies Valley north of Burns. The PLS built a big frame house in Harney City to serve as ranch headquarters near the planned county courthouse there, but when, after a local political struggle, Burns became the county seat, the company in December 1896 laboriously loaded the two-story building onto a freighter wagon drawn by 40 horses and mules, moving it ten miles over the icy high desert roads to Burns. The PLS headquarters remained there until Superintendent John Gilcrest moved out to the Alvord Ranch after John Devine's death. By then, gradually improving back-country roads and the coming of telephone service helped relieve some of the extreme isolation of the country.[5] The automotive revolution arrived when a Burns businessman opened the county's first automobile sales agency a few years later, and trucks gradually began to replace horse-drawn wagons.

Though sometimes resented for its size and power, the PLS was generally respected by locals for its capable management. In the early years of the century the PLS prospered under the diplomatic Gilcrest, a former surveyor and land cruiser whose talents Henry Miller had recognized when he was working on the Buena Vista Dam in the southern San Joaquin Valley. As PLS general superintendent from 1890 to 1920, Gilcrest took pains to implement Miller's policy regarding friendly relationships with neighbors and gave employment to many men who later went on to independent careers but retained a feeling of pride at having been associated with the company. Following Miller's own policy established in the San Joaquin Valley, Gilcrest quietly let it be known that any settler who truly needed food for his family might take a PLS steer from the range as long as there was no waste in slaughtering and the hide was hung in clear sight. (Some county residents later claimed that a number of local children grew up eating PLS beef.)

The PLS was frequently in court, however, over issues in both Oregon and Nevada, especially in cases hinging on water rights. Miller's attorneys had to be well versed in the varying details of water law in both states as well as in California. Miller himself possessed an instinctively legalistic mind and was adamant about pursuing arguments based on the meticulous interpretation of local circumstance as applied to law.

One PLS lawsuit reversed the arguments used by Miller & Lux in the famous Lux v. Haggin case by pitting the Nevada doctrine endorsing appropriation against the riparian rights of Tom Rickey in California. In this case the PLS owned land along the Walker River in Nevada and had water rights under Nevada law, but Rickey had land along the upper reaches of the river where it originated in the mountains of California in the Antelope and Bridgeport valleys. Besides taking water from the river to irrigate, Rickey conceived a plan to divert water into a reservoir and construct a tunnel to irrigate even larger acreages. On the Nevada side of the border, these activities began to diminish the river's flow in the downstream reaches owned by the PLS and others. Miller & Lux took the case to court, including their neighbors in the suit, to claim their water rights under the laws of Nevada.

As described by Edward Treadwell, then legal counsel for the firm, Rickey argued that his riparian rights in California trumped downstream users' appropriative rights in Nevada. The Rickey case became very convoluted, partly because of the matter of legal jurisdiction–whether California or Nevada water law should prevail–and went all the way through the federal court system up to the U.S. Supreme Court. The case was finally decided in favor of the Pacific Live Stock Company, but before the exact rights could be determined in detail for all parties (based on the thousands of pages of depositions Miller had earlier insisted on collecting and filing), Tom Rickey was indicted in California for financial fraud in another matter. He was forced to sell his Walker River property. Always the negotiator, Henry Miller made a settlement with the new owners of the Rickey property, arranging for the construction of a stream-fed reservoir to serve all claimants equitably.

PLS attorneys, including Treadwell who would serve for 15 years as general counsel for Miller & Lux, were kept busy with continuing lawsuits in Nevada and Oregon. Thus Miller & Lux contributed to the growing library of case law in three states.

1 Treadwell, *Cattle King*, pp. 116–47; other sources on the PLS include Simpson, *The Community of Cattlemen*, and Brimlow, *Harney County, Oregon, and Its Range Land*.

2 Treadwell, *Cattle King*, pp. 95–115.

3 The main source for the story of the P Ranch is *Cattle Country of Peter French*, but Edward Gray provides much supplemental detail in *Life and Death of Peter French*.

4 French, *Cattle Country*, p. 147.

5 Gilcrest, "Alvord Ranch Interlude," *Oregon Historical Quarterly* 1962.

~10~

End of an Era

CALIFORNIA'S WHEAT BONANZA lasted barely a generation. The great wheat harvests in the Central Valley began to decline in the late 1880s as the golden years of high market prices sloped gradually downwards. A broker's attempt to "corner" the California wheat crop in 1887 resulted in a price crash that sank some ten million dollars in investments, a huge amount in those times, and the market never recovered its former ebullience. Wheat exports fell rapidly after 1890. "The farmer looks back with regret to those 'good old times' and wonders whether he will ever again get two cents a pound for his wheat, as he did more than half the time from July, 1870, to July, 1878," wrote Horace Davis in a retrospective piece in 1898. "Indeed, it was too good to last." [1] The changing dynamics of the international wheat markets continued to lower prices for California wheat from year to year, until in 1894–95 the average had dropped from the profitable two dollars a cental of the late 1870s to about ninety cents, and, Davis reported, "In October 1894 it touched bottom at the disastrous price of eighty cents, far below cost of production," even with greater efficiencies in cultivation, harvesting, and transportation. From the state's peak production years of 1880–84, when an annual average of more than 25 million centals was recorded, the wheat harvest fell to an average of 16 million centals or less in 1894–97—a drop of well over 30 percent.

California growers cut back on grain acreage in the 1890s for a variety of reasons: because the patterns of the wheat trade had shifted in response to emerging global competition; because soils were exhausted after years of continuous cropping; because the big bonanza farms were being subdivided; and because new developments in horticulture gave farmers hope for more profitable crops. The lower market prices of the 1890s converged with high marketing commissions and freight rates to pinch farmers financially, but declining yields on the grain lands made matters worse. Grain growers had forced crops from their soils for years with meager attempts to conserve resources, and even regular field fallowing failed to restore the original fertility of what had been virgin land. In these years before California's variable soils were well understood, few farmers were able to replenish their fading nutritional values, and the development of specialized fertilizers lay far in the future. Fickle weather took its toll on harvests too, but now the poor crop years were no longer redeemed by the bumper harvests of other seasons. In the steady downward trajectory of the state's grain industry, some farmers depending on wheat were simply squeezed out of business.

The great wheat ranches began to be subdivided, and new settlers changed the face of the Central Valley. The newcomers, many of them attracted by railroad publicity, brought trade and growth to the new towns organized at convenient junctions. Some settled in the small-parcel "fruit colonies" that dotted both the Sacramento and the San Joaquin valleys, meeting with varying success. Improved water engineering offered hope for better irrigation systems, while advancing transportation networks and the development of refrigerated rail cars encouraged the marketing of more perishable crops. Farmers with good land and sufficient water began to concentrate on higher-value plantings, converting former grain fields to field crops, orchards, and vineyards. As the potential of areas beyond the river became more apparent, portions of the Sacramento Valley experienced a land boom. In sparsely populated Colusa County, where wheat had been the staple for a quarter of a century, irrigation only slowly made headway, but its promise encouraged the split of several large properties, including the Jacinto ranch.

After 1900 total state wheat production dropped precipitously, and by 1908 California was shipping less than 3 million centals of wheat annually. Wheat would remain an important commodity in many areas, including the Sacramento Valley, but it would seldom again bring the generous returns of the bonanza years, nor would it blanket hundreds of thousands of acres to the exclusion of other crops.

Hugh Glenn's widow Nancy died in San Francisco in October 1891. The Glenn holdings in the Sacramento Valley then passed to her three surviving children, now in their early thirties. None of them showed the interest of their father in active farming, and the properties were professionally managed by F. C. Lusk, an attorney in Chico, who took care of leasing land, selling grain, and other financial matters. Having grown up with a sense of entitlement, the three heirs paid relatively little attention to the management of the big estate except for the income it brought in, although they maintained the big house at Jacinto for a time. Lessors managed most of the details of production, but as wheat prices declined, so did the profits, even as managerial costs grew. Advised by Lusk, the Glenn heirs began gradually to liquidate their holdings as opportunity suggested.

With the weight of his father's name behind him, Frank B. Glenn, the younger surviving son, took part in local politics, serving one term as a state assemblyman in 1899–1900. He built a showplace country home called Glennair, surrounded by gardens along the river road a couple of miles above Jacinto. His older brother Charles, who owned and managed the Fouts Springs resort in the Coast Range to the west, maintained another home a little north of Glennair. Ella Glenn and her second husband, Charles Leonard, spent much of their time in San Francisco but stayed some of the time in the big house at Jacinto.

In 1900 much of the Glenn estate was offered for public sale—43,500 acres at $10 to $35 per acre in quarter sections. Sales proceeded gradually. In 1905 the family was still advertising 40- to 640-acre parcels in farm journals and newspapers. The final division of the estate took place in 1908, and that year the Alfalfa Farms Company, headquartered in San Francisco and now the owner of a significant portion of Glenn's former 86-square-mile principality, offered investors more than a thousand acres in 10 to 40-acre parcels pre-planted with alfalfa, a crop dependent on irrigation but more remunerative than wheat. The following year the Glenn heirs sold much of their remaining property to the Sacramento Valley Irrigation Company, which subdivided the land into 40-acre parcels and widely publicized their availability. Thus Hugh Glenn's great wheat ranch, which thirty years before had attracted national attention, receded into memory.

The great Chicago World's Fair of 1893 touted the promise of industrial progress across America even as the national economy took an abrupt downturn, the worst in its history until then. California suffered in some ways even more than other states. The Panic of 1893, precipitated in part by overproduction of silver in the western mines and then by a run on the gold supply, saw a series of bank failures and company bankruptcies that devolved into widespread unemployment, followed by a wave of labor strikes and an economic depression that hit the working class particularly hard. For nearly five years the nation struggled through financial hard times, worst in the cities but infecting farming areas as well. The unemployment level reached and exceeded a sobering ten percent. Mines closed throughout the West, some permanently, and in California the dominance of the Southern Pacific Railroad over shipping (and politics), the bursting of numerous speculative land and irrigation bubbles, and the long downslide of the state's wheat industry were just three of the disturbing factors in the troubled economy.

The murder of Peter French in 1897 was a potent reminder of the conflict not only between cattle barons and settlers but between owners and workers across the country. Yet the ferment of popular discontent in the 1890s expressed itself in different ways. Agrarian frustration was channeled into the Grange movement, while urban discontent helped fuel the rise of the Populist Party. Several political leaders, labor strikes, and demonstrations roused the people and made national headlines; but other efforts by government and private organizations began to meet needs in ways that would have long-term positive effects.

One highly effective and enduring effort to help farmers was embodied in the federal Hatch Act of 1887, which created permanent agricultural experiment stations within the states' land-grant universities. The University of California at Berkeley had already begun experimental work in the 1870s under the direction of Professor Eugene Hilgard, an eminent soil scientist who applied himself to the systematic mapping of California soils and the collection of information from practicing farmers across the state. Much of this wide-ranging information was published in the annual reports of the College of Agriculture, but in 1882 Hilgard used it to prepare a voluminous and detailed Report for the Federal Commissioner of Agriculture that became a valuable sourcebook on California conditions. Hilgard's work and that of other university scientists also received widespread publicity in the 1890s through the pages of the popular *Pacific Rural* Press. His research on soils was particularly important for farmers confronted with black alkali problems in the San Joaquin Valley, but he and his colleagues worked on a broad variety of issues. Hilgard gained international recognition when he received the prestigious Liebig medal in Germany in 1894 for advances in agricultural science. Encouraged by the Legislature, he also conducted experiments in winemaking, presaging the later work that would help develop the billion-dollar California wine industry. The College of Agriculture also responded to requests for courses on dairy production and extended new scientific findings through classes for local farm groups from Redding to Los Angeles. Thus government-sponsored research and outreach began providing technical assistance for California's growing number of farmers.[2]

The growth of a conservation ethic also began in the 1890s, starting with efforts along the Atlantic seaboard but best embodied in California by the founding of the Sierra Club in 1892 by John Muir and others. Muir had moved to California in 1868 and fallen in love with the mountains and Yosemite Valley. After his marriage into the family of a prosperous farmer near Martinez, Muir spent some years farming while continuing to write eloquent essays about the state's natural assets. His articles made him nationally known, and he became a leader in the emerging environmental movement that lobbied for the establishment of national parks, forests, and preserves. Though he was instrumental in gaining national park status for Yosemite Valley, he failed in his efforts to prevent the conversion of the nearby Hetch Hetchy Valley into a reservoir for urban water supplies.

Livestock's devastating impact on the land provided much of the rationale for growing western environmental activism in the 1890s. The previous decade had seen the cattle industry under mounting pressure from complaints about abuse of public lands. Some cattlemen were fencing tracts illegally, and western ranges were suffering visibly from overgrazing. Although the Department of the Interior had ordered unlawful enclosures to be removed, few cattlemen complied until Congress moved decisively to reform western land policy. After repealing the Preemption Act of 1841 and the Timber Culture Act of 1873 and tightening the Desert Land Act provisions, Congress went a step further in 1890, approving the creation of Yosemite National Park, General Grant National Park, and the Sequoia Big Trees Reservation. Livestock grazing was thereby severely restricted within large public areas in California. Added to the restrictions a few years later were several large forest reserves created in Southern California and the more than four-million-acre Sierra Forest Reserve covering parts of Mono, Mariposa, Fresno, Tulare, Inyo, and Kern counties.[3]

The range cattle industry in California was now truly at a turning point. Cattlemen accustomed to free range feeding either had to go out of business altogether or pursue stock farming like Miller & Lux. To meet the challenges cattle and sheep ranchers formed the American National Livestock Association in 1897, and later the Stockmen's Protective Association, to promote their own ideas on national and state policies. Continuing controversy over the use of public lands led to the institution of a federal grazing permit system in 1901, under which the Department of the Interior would issue permits for a certain number of stock on a given area for a certain length of time. Later the responsibility for grazing permits was transferred to the Department of Agriculture, which began imposing modest fees. The debate over livestock grazing on public land continued through the presidency of Theodore Roosevelt, with the federal government under continuing pressure from environmentalists to take greater action to conserve dwindling resources.

The changes in livestock production paralleled the growth of the human population. Between 1880 and 1910 California's cattle increased from under half a million to 1.6 million head—a tripling of production in 30 years. The 1880s had seen a deficiency in beef supply relative to the population; nearly a third of the animals slaughtered in California in 1885 came from Nevada or Oregon. Beef prices fluctuated in response to the limited supplies until greater stock numbers from more producers in the 1890s led to a steadier market. The extension of railroad lines helped shipping, and bigger stockyards fed more animals. In 1902 the State Agricultural Society published a report calling these cumulative changes a "revolution," "a new and better way" to raise beef for market by keeping smaller herds on cultivated pastures and producing better, heavier animals. One scholar has defined these changes as the industrialization of cattle production.[4]

Miller & Lux and other big cattle firms were on the cutting edge of these changes. Mixing traditional range feeding with far more use of cultivated hay and alfalfa, their livestock operations were integrated into general farming operations. Cattle spent only the spring months in the foothills, fattening in lowland pastures and feed yards the rest of the year. By 1893 Miller & Lux ranches were trough-feeding steers up to four pounds of grain and 25 of hay daily, supplementing this diet with oil cake, sugar beet pulp, even pumpkins in autumn. Although the company did not concentrate on purebreds, it continued to upgrade its herds with shorthorns, Durhams, Devons and mixed breeds for various purposes. One of the ranches also began breeding heavy work horses for use on other ranches and for the San Francisco Fire Department and drayage firms in the city.

Competition for water in the San Joaquin Valley remained a fact of life. Henry Miller had always sought incontestable water rights for his holdings, but in his pragmatic way he took a broad view over the settlement that continued to redefine the valley. The San Joaquin & Kern River Canal & Irrigation Company, controlled by Miller & Lux, became a public utility as well as an irrigation canal, with customers in three counties. In the 1890s the company extended the Dos Palos Colony lateral canals and completed the Outside Canal, and several new creameries began to serve a growing number of dairy farmers. Gradually, under court order to sell assets to satisfy Lux's German heirs, Miller began to release some of the acreage he had taken so many years to acquire. During the decade Miller & Lux sold 4,000 acres in 20- and 40-acre parcels to settlers in the Dos Palos Colony; and then sold 10,000 more after 1900, allowing 40-acre farms to fan out around the town of Gustine, founded in 1907 along the Southern Pacific line (and allegedly named in memory of Miller's deceased daughter, nicknamed "Gussie"). Through these relatively small reductions in the firm's holdings, hundreds of settlers were now able to pursue new lives.

Despite Miller's improvements in canal operations, the firm's control of water rankled some neighbors, for smaller landowners paid up to twice the water delivery rate of Miller & Lux. In 1896, after unpaid bills were followed by water cutoffs, members of a newly organized water users association pressured the Stanislaus County Board of Supervisors to hold hearings. As a result the SF&KRC&IC lowered rates for those growing thirsty crops like alfalfa, but litigation over "surplus" water and new contracts continued, and Miller kept his attorneys busy reviewing the cases. By 1900 the two companies—cattle firm and canal company—were virtually one, Miller serving as president of both, and administrative matters were handled by a combined staff. The Miller & Lux ranches continued to receive water at less cost than others, and although for a time this practice was supported by court decisions, the "tangled thicket of lawsuits, countersuits, and appeals" became expensive and time-consuming. Until well into the twentieth century, Miller & Lux litigation on these and other water matters continued to occupy the courts.[5] As Miller himself remarked, "I have made three fortunes—one for myself, one for my partner, and one for my attorneys."

Beyond the rangeland and the pastures, the ranches and the farms, the markets for which valley cattle were destined became a new battleground. Ironically, the former Charles Lux estate south of San Francisco, where herds had historically been held before being sent to Butchertown, became the scene for a hard-fought "meat war" that pitted local butchers against the giant meat packers from Chicago.

By the mid-1880s most of the wholesale meat business in the Midwest had been consolidated by the "Big Four" meat packers of Chicago—Swift, Armour, Cudahy, and Morris—who began to take steps to expand into the rest of the country. Gustavus F. Swift was the leader in this drive. He arranged to build strategically situated stockyards and packing plants all the way from Minnesota to Texas, and developed refrigerated rail service to ship dressed beef from these plants. As Swift and his colleagues began extending their reach toward the east, however, threatening the trade of local butchers and retailers there, their opponents organized the National Butchers' Protective Association to encourage boycotts against the Chicago companies. These efforts slowed but did not stop the inexorable march of the big packers across the national stage. Swift soon set his sights on the San Francisco Bay Area and its trade with the entire Pacific Coast, making plans to establish a joint stockyard for several packing plants to be operated by Chicago firms, which could then expand their sales into a whole new part of the country.

The family of Charles Lux put his estate at Baden up for sale shortly after his death in 1887–convenient timing for Swift and his partners, who recognized the advantages of the site (now part of Brisbane) for their purposes and took steps to acquire the property along with other parcels nearby. In 1891 the Chicago syndicate with other backers purchased 3,500 acres for $750,000 near Point San Bruno, meanwhile organizing the South San Francisco Land and Improvement Company. Plans were laid for a new city to be called South San Francisco, analogous to South Chicago and other areas for which Swift had masterminded development. This site along San Francisco Bay offered an excellent location for meat packing facilities; water and rail transportation were already available, and prevailing winds could blow the odors from the stockyard toward the bay. Employees would live in the new town to be developed to the west.

The meat packing conglomerate was already shouldering its way relentlessly into markets across the country, and with such careful planning the time seemed ripe for a new success in the Far West. San Francisco meat vendors, however, rose up to fight the plan with vigor. In 1893 the Bay Area meat wholesalers, led by Henry Miller, organized the Butchers' Board of Trade (BBT) of San Francisco and Alameda Counties. At the same time Miller, always the realist, hedged his bets by investing $40,000 in the South San Francisco Land and Improvement Company, taking a place on the first board of directors. In this strategic position he would be privy to the plans of the Chicago packers and might also serve as an intermediary between the syndicate and the BBT. With a foot in both camps, he would also be unlikely to lose business.

The BBT soon organized a retail boycott against its chief new competitor, the Western Meat Company owned by Swift. The ensuing "meat war" lasted more than a year. Faced with stiff opposition, nearly all the other Chicago investors began to have second thoughts, and the financial crisis of 1893 discouraged their plans as well. Some withdrew from the scheme. Only Swift's company in South San Francisco managed to survive by relentlessly slashing prices and opening its own store. Although Western Meat was able to win several large contracts to supply meat to the state prisons and to the military, the BBT retail boycott generally remained effective until the 1906 earthquake destroyed old Butchertown. Meat marketing in San Francisco would then move into a new era.[6]

Widely recognized and respected for his accomplishments, Henry Miller had become something of an elder statesman by the 1890s. Even as the economy evolved in new directions, Miller was still firmly in command of his business as he approached the age of 70. Complications arose, however, when the seven-year Miller & Lux partnership agreement following Lux's death came to an end in 1894. That year the "German heirs" appeared to claim their inheritance. Miller had rarely allowed money to be taken out of the business after his partner's death, claiming that nearly all the profits were needed for maintenance and improvement of operations on the company's far-flung properties, or at the facilities in Butchertown where the trade was growing and needed more cold storage. Since no one knew as well as he what was going on in the business at any given time, no one had been able to gainsay him.

Miranda Lux died in 1894, and her son Jesse Potter and the other heirs were impatient to settle the books at last, filing suit to do so. The court appointed three commissioners to arbitrate the division of Miller & Lux assets. Several years of patient and impatient negotiating ensued. More lawsuits were filed, a revolving cadre of attorneys appeared in courthouses, and animosity built on both sides. At one point some of the Lux heirs petitioned for the removal of Potter as executor on grounds of "incompetence, improvidence, and conspiracy," but they later dropped this effort and redoubled their attacks on Miller. Local newspapers gleefully reported some of the more colorfully malicious arguments. Finally, in 1897, Miller agreed to reorganize and incorporate the original partnership. The Lux heirs were issued stock certificates for the value of their claims. Miller then methodically began to buy all the outstanding shares. In 1901 the Superior Court in Redwood City declared the case closed after years of litigation, but it took another decade to complete transfers of the remaining stock.

Meanwhile, Miller hired a young European with a complicated name to serve as superintendent of the headquarters Santa Rita ranch at Los Banos in the late 1890s. Henrik Albert Van Coenen Torchiana was a well-educated, dapper young man of good family, whose taste for adventure brought him to California in his early twenties. After trying unsuccessfully to establish a citrus orchard in Southern California, he took a job on a small ranch in the San Joaquin Valley, where Henry Miller noticed his abilities and hired him away to work at Santa Rita. There Miller tested him in different positions and quickly moved him up in the hierarchy to serve as the Santa Rita superintendent.

Torchiana was superintendent for some five years. Eventually he was ready for a change, not least because friction had begun to develop between him and the "Old Man." Torchiana, who had been studying law in the evenings at Santa Rita, left the ranch, passed the bar, and began practicing law in Santa Cruz. He moved to San Francisco to continue his career, and in the 1920s he was appointed consul general for the Netherlands. In 1930, well after Miller's death, Torchiana published a fictionalized account of some of his experiences working for Miller, entitled *California Gringos.* This thinly veiled memoir celebrated the romance of the West with himself as the main character, while painting a picture of the legendary cattleman at the turn of the century.

"When is a cowman not a cowman?" Torchiana asked in his preface, launching into a soaring description: "There can be only one answer....Never. For he who has once breathed the air laden with a sweet odor of blossoming sage-brush; who has seen the stars slowly vanish from the canopy of heaven while he stretched himself luxuriously in the blankets, just prior to greeting the dawn with a song–or a mouth-filling oath; or has felt the thrill of victory course wildly through his blood at the tightening of the riata, the bracing of the horse and the dull thud of the fleeing animal on the turf; can never forget that he was at one glorious time a centaur of the plains and the mountains." [7]

Though *California Gringos* largely reflects its author's nostalgia for the past, it also pays tribute to Miller, called by the pseudonym "Mr. Carpenter": "Of thick-throated speech, flat skull, stocky powerful build, at seventy years of age his carriage was erect and he walked with an athletic stride. Self made, amassing a huge fortune in land and livestock, due to great ability and unflagging industry, he had found time to acquire only a few of the gentler graces...[with].a masterful mind...an iron will...and endowed with a tremendous vitality, nature had bestowed on him a surplus of brain and brawn, leaving him unburdened with a ballast of too finely spun ethical feelings....a man of boundless energy who had never spared himself, any more than anyone else."

Clearly impressed with Carpenter's intelligence, Torchiana calls him a genius more than once: "By nature suspicious of all modern book learning [yet] he had the political and legal history of the State of his adoption at his finger tips....his mind was a human calculating machine....He had a genius for land law. The homestead, timber claim, mining claim, school land laws were all at his fingers' end....he was better posted on riparian and other irrigation law than most lawyers. A genius in organizing this great, complicated, widely flung industrial-agricultural plant....He never carried even a pocket memorandum, but depended entirely on his marvelous memory.... having acquired his vast knowledge of matters agricultural gradually as the properties grew; and his great success being based on practical experience and close observation.....[he was a] genius under whose hands a whole countryside had blossomed into prosperity.... this Napoleon of the western plains." [8]

The portrait that emerges in *California Gringos* is truly that of an aging Napoleon–harsh and autocratic, but beginning to be vulnerable, as in the scene where he falls while crossing a narrow walkway over an irrigation ditch and breaks a couple of ribs, injuring himself painfully (a true story). Time is no longer on Miller's side, and the world around him is changing. Torchiana knows that his own departure from the ranch is inevitable, yet leaves his boss, the old warrior, with mixed feelings.

Anecdotes and stories grew up around Miller as he aged, reflecting the mixed views of his associates and employees. Some thought Miller penny-pinching, while others admired his generosity. Many anecdotes describe his frugal instructions to subordinates, while others recount his acts of unsolicited kindness. The stories of his insistence on thrift elicited a kind of head-shaking ambivalence. He was said to have told a foreman to instruct his men to pull cockleburs while they were waiting in the fields for ewes to lamb, so that they would not waste time; a young girl sweeping a porch to store her broom upside down on the handle so that the bristles would not wear out so fast; and a ranch manager complaining of mice running through the buildings to get cats and then send their kittens on to the next ranch when the mice were under control.

Then there were the stories about robberies. In one popular anecdote, Miller was stopped by a highwayman while he was traveling alone in an isolated area. As he was being relieved of his cash, he asked the robber to let him keep $20 to get back to headquarters. The man gave him $20 and took the rest. Some months later Miller happened to see the highwayman in a saloon, walked over to him and handed him $20, remarking that he was repaying his debt. This story suggests that Miller was not entirely humorless—but it also reveals his view of thieves, for Miller was not infrequently targeted for robbery as he traveled through the valley over the years. He took care to carry with him only the amount of money he could afford to lose, and he never pressed charges, probably to avoid cause for retribution. In a time when rural violence was not uncommon, no robber ever harmed him.

Economy was always on Miller's mind, yet as he aged he could afford to be generous, and he was. He founded the May Day festival at Los Banos in 1893, an event that became an annual tradition. The company donated meat and other provisions for the community barbecue, and Miller himself for some years helped serve the crowd. (When asked to address the gathering, however, he always refused, for he never liked public speaking, perhaps because of his heavy accent.)

Other stories of Miller's generosity in his later years were legion. Himself a self-taught man who knew the disadvantages of his youth, Miller paid for the education of 15 nieces and nephews as well as the children of numerous employees. At Christmas he gave away hundreds of turkeys grown on the ranches for the holidays. He sometimes gave retired employees the deeds to their houses. He

Henry Miller established the tradition of the May Day celebration in Los Banos in the 1890s. He served beef from the Santa Rita Ranch personally to community members at this event.
Courtesy of Milliken Museum, Los Banos.

made large donations to the German hospital in San Francisco and to other institutions. He gave many gifts anonymously and forgave debts when he thought the debtors worthy. He was generous to beggars, although he abhorred drunks. Hoboes often lined up outside his San Francisco office when they knew he was there, knowing that he would hand out money to all when he came out the door.

Miller's cattle empire continued to dominate a large part of the San Joaquin Valley and much of the Oregon/Nevada corner of the inland plateau through the early twentieth century. When other landowners sold out to speculators, he stuck to the cattle business. When oil was discovered in the area around Bakersfield, Miller allowed some of his land to be leased by oil companies that needed access to water, but he never personally participated in oil drilling schemes. Even as the oil boom began to re-shape the economy of the southern San Joaquin Valley and some of his relatives acquired promising property, Miller himself remained profoundly conservative in his use of land.

Other parts of the San Joaquin gradually converted from livestock and grain production to more intensive and profitable crops made possible by irrigation and improved transportation. Not all ventures were successful. Some of the early colonies failed as irrigation began to increase the alkalinity of local soils, degrading them for cultivation. Other ambitious schemes were thwarted by bankruptcy or death, like those of Theodore Kearney.

Kearney had begun to promote raisin growing in the 1890s as an important new crop for Fresno County and for the colonists he was trying to attract to his property called Fruitvale. He helped organize a raisin marketing cooperative, which despite some setbacks became one of the state's first successful commodity marketing associations. Kearney also made plans to build a grand estate west of Fresno, with a large French-style chateau to be located at the end of a long line of palm trees. His bold plans ended with his unexpected death at sea on his way to Europe in 1906, the chateau still only a vision (though the palm trees remain). Kearney's influence lingers in the boost he gave to the raisin industry by promoting cooperative marketing.[9]

Claus Spreckels also helped shape a more diversified agriculture for California, and his reach extended beyond the Pacific Coast. He had started with sugar production in the 1860s, using cane sugar from the Hawaiian Islands. By the 1880s the "Sugar King" branched out into ventures in transportation, banking, and real estate, becoming heavily involved in shipping through his Oceanic Steamship Company. He also constructed a massive irrigation system for his investments in Hawaiian farmland, a 45-mile canal delivering water to a 3,000-acre sugar cane plantation and four sugar mills.[10]

But it was Spreckels' competition with the Sugar Trust in America that propelled him into sugar beet growing and made him both a California agriculturalist and a nationally known capitalist. When the American Sugar Company based in the East tried to invade Spreckels' territory in the West, he fought back by competing with better refining technology. After purchasing large tracts of land near the California coast for the cultivation of recently introduced sugar beets, Spreckels went to Germany in 1887 to study the beet sugar industry there. He worked six weeks as a laborer in a beet sugar mill at Magdeburg to learn the business from the ground up, then ordered German machinery for his new refinery at Watsonville in the Pajaro Valley. The first plant of the newly renamed Western Beet Sugar Company began operations in 1888 and was soon served by a 42-mile narrow-gauge Pajaro Valley Railroad connecting the Salinas Valley beet fields with Watsonville. Thereafter California beet sugar production expanded rapidly. Spreckels announced plans to build a Philadelphia refinery to compete with the Sugar Trust, but the Trust capitulated to this threat by agreeing to buy that refinery in 1892. When Spreckels began construction of the world's largest beet sugar factory in Salinas, the Trust arranged to buy half that facility too, which opened in 1899.[11]

Claus Spreckels took on another giant in the 1890s by taking the lead in organizing and building a new rail line down through the San Joaquin Valley—the San Francisco and San Joaquin Valley Railway—to compete with the Southern Pacific. This new line was welcomed by local shippers who had felt squeezed by high SP rates. Spreckels, the major investor, served as president of the SJVR from 1895 until it was sold in 1901 to the Santa Fe Railroad.[12]

Henry Miller invested in the company even though he was known to dislike Spreckels, taking a seat on the railway's board of directors. He turned the construction of the rail line to his advantage by negotiating the right-of-way through Miller & Lux overflow land. When the SF&SJVR was forced to abandon an already completed section, Miller acquired 100 miles of the railroad bed free, a raised grade that became a convenient levee road for farm operations.

In 1905 the *New York Times* listed Claus Spreckels as one of the ten richest men in America. Not only fabulously wealthy as the "Sugar King," with millions in equity in sugar refineries, shipping, and farmland, he was also a city real estate developer with investments in San Francisco office building, banks, and utility companies. The Spreckels family lived in a grand three-story stone residence on Van Ness Avenue until it burned in the fire after the 1906 earthquake. But Spreckels also owned a splendid house on New York's Fifth Avenue, where that city's social pages noted the coming and going of family members on their way to Europe.

Young California novelist Frank Norris was fascinated with both the romance of wheat and the dominating role of the railroad in California life. In the late 1890s he spent two months on the Santa Anita ranch near Hollister (a Miller & Lux property) preparing to write *The Octopus*. The novel dramatizes (and exaggerates) the role of the railroad in strangling the state's economy, while following the progress of the wheat crop over the course of a year. Norris turned the real town of Tulare into the fictitious town of Bonneville and used the Mussel Slough incident as the pivotal event in his theatrical re-writing of San Joaquin history. The novel was published in

1901 to popular acclaim, and Norris planned to write two sequels about wheat marketing and international distribution. He died from pneumonia in 1903, however, before the second novel (*The Pit*) was finished, and before the full maturing of his talent. Norris was only 32 when he died, and his writing reflects an evolving rather than a settled social viewpoint; but in many ways his work is representative of Californians at the turn of the century–nostalgic for the color of the past, awed by the natural beauty of the landscape, eager for settlers to make good and angry at monopolists, simplistic in understanding of economic problems but feeling their way forward toward a more open and just society.

Much was happening on the national and state levels that reflected those sentiments. California agriculture was moving into a whole new era. In 1902 Congress passed the national Reclamation Act with the aim of encouraging small farms in irrigated areas, and a swell of interest in the law's potential for California lifted many expectations. The Bureau of Reclamation assigned Elwood Mead, an engineer known in Wyoming for his contributions to water policy, to study the implications of the legislation for farming in California. Mead's report, *Irrigation Institutions in California,* focused on several functioning irrigation systems with a view toward developing others that might help small farms thrive. At the same time the University's College of Agriculture was expanding its offerings to meet the demand for practical knowledge as well as scientific research. In 1905 the Legislature established a State University Farm, soon sited at Davis, where farmers could attend short courses and their sons could get a campus-based agricultural education. Growing public interest in agriculture was also reflected in press coverage and in popular lecture series sponsored by such groups as the Commonwealth Club in San Francisco.

Adding to the transformation of rural areas was a new group of immigrants from Asia. Although federal law now restricted immigration from China, no such laws yet applied to other Asians, and they came in growing numbers. Some came from northern India; Punjabis appeared as farm workers in Southern California around 1900, migrating north over time. Some came from Japan, and like the Chinese they were very good at labor-intensive farming, at first working as contract crews but gradually starting their own enterprises in truck farming, nurseries, and gardening, especially in Southern California but also in the Central Valley and the Delta. Japanese farmers were patient, careful, and persistent on marginal soils, and not lacking in ambition. They made their way forward even as apprehension grew about their competition.

Kinji Ushijima, better known by his American name George Shima. became California's best known Japanese farmer. A well-educated man who came to America in his 20s in 1888 to learn English, he decided to stay permanently and began to work as a farm laborer. He soon became a labor contractor and lessor of land in the Delta around Stockton, and over a period of years became famously successful as a potato grower on an estimated 10,000 acres, utilizing a fleet of steamboats and barges to take his crop to the San Francisco market. His acreage included what is today known as the Shima Tract near Stockton, an island that he reclaimed with the help of hundreds of Japanese laborers. He was also one of the first to grow asparagus in that area.[13]

Many Americans did not welcome the success of Japanese immigrants. Anti-Japanese prejudice brought about the so-called "Gentlemen's Agreement" in 1907, in which the U.S. and Japan agreed to restrict further immigration but to permit family members to join those immigrants already here. Shima, a cultured man with strong connections to the Japanese Embassy, founded California's

Japanese Association in 1908 to encourage mutual aid among the immigrant families.

The great San Francisco earthquake and fire of April 18, 1906, was a turning point for California. It devastated the most important city in the state, leaving entire neighborhoods in ruins and, for a time, many businesses in shambles. It would take years for San Francisco to recover, but rebuilding efforts would make the city proud again.

Henry Miller was at home on Rincon Hill with his invalid son when the earthquake hit. The lower slopes of the hill had grown crowded with small, poorly built wooden houses, and many of the larger homes had become boardinghouses, but Miller had kept his big house at 34 Essex as the family domicile. His wife Sarah had died after years of poor health, and Miller, now in his late 70s, was feeling his age. As the flames crept toward his home Miller arranged for his son to be bundled into a wagon, and then he himself carefully locked the front door, hoping to catch the train down the peninsula to Bloomfield. Instead the two spent the night in a cheap hotel in Butchertown while the fire raged through the city. The Miller house was consumed in the flames, little remaining of the family's thirty years there but the door key in Miller's pocket. The next day, as father and son fled, the fire reached Butchertown, the old wooden slaughterhouse buildings on the wharf falling into the water in jumbles of charred boards. Weeks later Miller moved back into San Francisco to live with his daughter's family in their home at 2101 Laguna, which had survived the disaster. The following year Henry Miller, Jr., died at age 44, leaving a widow but no children.

In his methodical way Henry Miller began to take steps to prepare for his own death and the transfer of the business to the next generation. In 1907 he hired Edward F. Treadwell as the company's first in-house counsel, primarily to take care of the always contentious water issues. Other attorneys were employed as needed, but Treadwell became the legal voice of the firm for the next fifteen years. Miller also resigned the presidency in favor of his son-in-law, J. Leroy Nickel. Nickel's former career as a liquor merchant had not prepared him for his new position as head of a land-and-livestock company, and he brought considerably less on-ground experience and a very different style to the job. He tried to modernize the company but almost immediately began to antagonize long-time employees and neighboring landowners as he tried to cut costs and reorganize long-standing work patterns. The increasing resentment did not help the company.

In 1910, now 83 years of age, Henry Miller made his last trip to inspect his ranches. He had his attorney prepare a Deed of Trust and took out a $5 million bond at the Bank of California to acquire all the remaining interests held by the heirs of Charles Lux. In 1913 Miller officially turned over all management duties to J. Leroy Nickel. By this time the old man was in a wheelchair, and the following years proved difficult for the firm and the family as he suffered from declining health. The world had changed, Miller was no longer in charge, and Nickel was no Henry Miller. The old cattleman spent his afternoons driving through Golden Gate Park in a carriage and managed to visit the livestock exhibits at the Pan Pacific Exposition of 1915, but was more or less confined to his daughter's home thereafter.

On October 14, 1916, aged 89 years of age, after lapsing into a two-day coma, Henry Miller died quietly at home. In accordance with his instructions, the family held only private services and arranged no public memorial event, and his body was laid to rest on his beloved estate at Mt. Madonna.

The death of the old titan—the hungry young German butcher's assistant who had come to California during the Gold Rush and built an empire from nothing–made the front pages of all the local newspapers. His obituary in the *San Francisco Call* noted that "the long trail is ended" for the "last cattle baron," eulogizing him as a man of "no sham, no pretense...a worker to the end." Among recollections elsewhere, the editor of the *Gilroy Advocate* wrote that Miller's unfulfilled ambition had been for the state to establish a new county on the west side of the San Joaquin Valley, to be known as Miller County with Los Banos as the county seat. The *Pacific Rural Press* printed a fine portrait of Miller in his old age along with a somewhat ambivalent editorial on his "long, masterful life," noting his insistence on control but expressing admiration for his clarity of purpose, his lifetime of perseverance and "joy in work," and calling him without reservation an executive genius.

Perhaps the finest eulogy of all came from John Clay, an aging western cattleman, a few years after Miller's death. In the early 1920s Clay published a memoir entitled *My Life on the Range*, which included a tribute to Miller that was reprinted in *The Cattle King* in 1930. "Nothing in my experience...equaled the matchless energies displayed on these San Joaquin ranches," wrote Clay. "The great motive power, the irresistible force in this ranch life and management was Henry Miller, a nervous little German who possessed some fairy's wand, for he transformed whatever he touched into gold. He was selfish, grasping, indomitable, thrifty, with a wondrous brain that schemed and twisted and generally routed his opponent. Yet he was a philosopher, a student of the Bible, because he drew illustrations from the life of Solomon....In his business he was decisive and you knew where you were at. ...his word was good when given. Aside from his own genius his great asset was in choosing men. He had an army of them and they were splendidly drilled, loyal and intense workers. His magnetism permeated the valley...from the Mexican vaquero to the boss of the ranch [they] worshipped him and would fight for his honor for they considered it their own. The love of your fellow man is the best monument that you can leave behind." [14]

Now began what would be a long struggle over the disposition of Henry Miller's assets. His daughter Nellie Sheldon Miller Nickel was Miller's only surviving child. According to his will most of the vast accumulated wealth of his estate–an estimated $20 million–was directed to pass in a trust through Nellie, whose three surviving children would be the primary heirs. (Henry Miller Nickel, the cattleman's second grandson, had died in 1909 at the age of 20 in a winter blizzard on the rugged Alvord desert near the Mann Lake Ranch in Harney County, Oregon.) But there were many other beneficiaries.

The will, a copy of which is preserved in the California Historical Society library, demonstrates Miller's typical attention to detail, in its numerous pages making generous bequests to the poor of Gilroy, the little town of Brackenheim in Germany, and literally scores of employees and others listed by name. Newspaper reporters, fascinated by the story and the fortune, noted that in his later years Miller had already given to many charities "without ostentation" and had planned to build an endowed hospital in Gilroy with a bequest of $300,000. He also made a bequest for $500,000 for medical research to be carried on at the University of California or at Stanford University.

At the time of Miller's death in 1916, federal tax authorities estimated his holdings at some 1.4 million acres along with control of far larger acreage through lease and grazing rights. He was reported to own more than one million acres in nineteen California counties, primarily in or bordering the San Joaquin Valley, and to control another two to six million acres in Oregon and Nevada. Soon came the battle over inheritance taxes. The state of California had already begun levying taxes on business, including an inheritance tax, but the federal inheritance tax was new, passed in 1914. A note in the *New York Times* of November 13, 1916, reported that Henry Miller had planned to avoid the federal tax by transferring his property to his heirs three years earlier, before the new law would take effect–strategic thinking up to the very end, but not in the long run successful.

The passing of the patriarch signaled the passing of an era in more ways than one. Even as the "Great War" made headlines in the newspapers, new technology was beginning to change daily life, and other prominent figures in California passed on around the same time. John Muir died at the end of 1914, closing a venerable career as one of the nation's great conservationists, and Eugene Hilgard, the renowned agricultural scientist at the University, died only a few months before Miller. Miller's death came the year after the city of San Francisco had exuberantly hosted the Pan Pacific Exposition to celebrate the city's rebirth after the great earthquake–and to observe the opening of the Panama Canal and the completion of the transcontinental Lincoln Highway, both of which promised dramatic improvements in long-distance transportation. Henry Ford's horseless carriages were going into mass production as the national road movement spread, and rapid sales of automobiles were matched with real estate development in the Bay Area. Not least, engineers were ushering in a new agricultural revolution with the growing availability of tractors.

The era of the nineteenth-century agricultural "kings" faded as the world changed. Henry Miller had spent a long and productive life acquiring land and water resources and enhancing the value of his holdings through careful management and aggressive litigation. He had earned fame and fortune for himself, but he had also helped open up the sprawling Central Valley for the prosperity of many others in the future.

---

1 Davis, "Wheat in California: A Retrospect and Prospect," *Overland Monthly*, 1898.

2 Scheuring, *Science and Service*, recounts the role of the University in California agriculture.

3 This section derives largely from Pulling, *History of California's Range Cattle Industry*.

4 Igler, *Industrial Cowboys*, pursues this theme.

5 See Miller, *Flooding the Courtrooms*.

6 See Igler, pp 160–167.

7 Torchiana, *California Gringos*, Preface.

8 Ibid., pp. 1–4.

9 Rehart and Patterson, *M. Theo Kearney: Prince of Fresno*.

10 See Adler, *The Sugar King in Hawaii*.

11 Cordray, *Claus Spreckels of California*.

12 *The San Francisco and San Joaquin Railroad*, a promotional book preserved in the California Historical Society Library, describes this enterprise.

13 Yoshimura, *George Shima, Potato King and Love of Chinese Classics*.

14 Clay, *My Life on the Range*, quoted in Treadwell, pp. 362–63.

Epilogue

After the Kings

THE STORIES OF HUGH GLENN AND HENRY MILLER are stories of a unique era. These men had the unmistakable advantage of being nearly the first on the scene during a wide-open period of opportunity, and their ambitions drove them to seize the day. Observant, capable, and bold, they confronted risks and overcame challenges that defeated men of lesser drive, but they were lucky too in their connections, which helped them acquire the capital needed to build large enterprises. They were also lucky to be able to withstand the natural or man-made disasters that wiped out many other men. The fortunate convergence of all these factors, however, could not last forever.

Hugh Glenn's famed career as the Wheat King followed the rapid upward trajectory of the great grain trade that flowed out of California in the years after the Civil War, but his death cut short his contributions to it at the very peak of the bonanza, and he did not live to see its lingering decline. With his leadership gone, the Jacinto Ranch began to slip into history as its holdings were sold and land development schemes began to transform the Sacramento Valley.

In 1908 Glenn's three surviving children agreed to a final division of the family's remaining property. In 1909, with erosion threatening the family graveyard along the river, Frank Glenn arranged to remove the remains buried there and send them to their final resting place in the Mountain View Cemetery Columbarium at Oakland. The next year the Jacinto post office, which had opened in 1858, was officially closed, ending the two-generation tradition of Glenn family service as postmaster. The willful Sacramento River changed course as weather varied over the years, meandering away from its earlier west bank, and the former landing at Jacinto lapsed into obscurity, its access to water made difficult as new sediment and vegetation redefined the channel. The former bustling village began to disappear, its functions shifted to other places and its residents leaving for other opportunities. Less than a ghost town, Jacinto receded into memory, only the footings of old warehouses today revealing where the river traffic of an earlier era took heavily laden grain barges down to San Francisco Bay.

Land sales continued to shrink the once-princely estate, and other farmers worked the land. On the broad mid-valley plains some of the early farm subdivision ventures languished as new owners found it difficult to make a living on small acreages in this area of challenging soils. Many parcels were eventually re-absorbed into larger holdings, especially as rice became the preferred crop on the heavy basin land. Some farmers planted orchards on the deep alluvial soils bordering the river, but extensive rather than intensive farming continues to dominate the area. Broad fields still stretch for miles toward the mountains bordering the valley on either side, their snowy peaks glistening on clear winter days. To the south of the former Glenn holdings, half a dozen state and national wildlife refuges now offer shelter to waterfowl on their migrations through the Pacific Flyway, presenting spectacular shows to dedicated birdwatchers.

The Glenn heirs shaped lives elsewhere, but retained a connection to Jacinto until their deaths, Charles in 1921 and Frank in

1930. Frank Glenn's home along the river was rebuilt after a fire, but the arched gate "Glennair" remains. Charles Glenn's summer resort at Fouts Springs, isolated at the end of a mountain road in the coast range, eventually became the site of a regional detention facility for troubled boys. Ella Glenn and her husband lived off and on in the big Glenn house near the Jacinto landing until his death, after which Ella lived quietly for periods in a Willows hotel until her son Harold's death in San Francisco in 1936. Inheriting the trust fund left to him by Peter French, she moved to a city residence club, where she died in 1938 at the age of 78, the last of Hugh Glenn's children. The Jacinto mansion, rented out for some years, deteriorated gradually until it stood abandoned and decrepit among its old palm trees along the river. The structure was razed in the 1960s, and only the overgrown foundation remains.[1] The former elegant Glenn home in Oakland acquired the reputation of being "jinxed" after three families living there suffered untimely deaths. Over the decades it too deteriorated into dilapidated multi-family housing in a declining neighborhood, until it was torn down in the 1950s to be replaced by an apartment building.

In Oregon's Harney County the legendary French-Glenn ranch under the ownership of the Blitzen Valley Land Company went through hard times and became the property of Swift and Company in the 1920s. During this difficult period for cattlemen the company made scant profit and finally sold the holdings to the federal government in 1935. The old P Ranch became the heart of today's big Malheur National Wildlife Refuge, a sprawling complex of managed fields and marshland that attracts millions of birds from the Pacific Flyway on their seasonal migrations. The original headquarters of the P Ranch can be viewed outside the hamlet of Frenchglen, where a small hotel provides lodging for refuge visitors. A few miles to the north, in the meadow where Peter French and his crew trained horses during the winter, his old round barn has been preserved as a local landmark. Ranching still dominates the local economy, but outdoor tourism now also provides income for those who live in this remote area.[2]

In contrast, Henry Miller's great cattle company outlived him for nearly half a century, though it was never the same without him. The company remained a regional rather than a national firm and began to decline in influence. In 1917 the Miller & Lux assets of $35 million placed it among the top 200 U.S. corporations of the time, yet despite its size and apparent wealth the company was not correspondingly profitable. The decline had begun even before Miller's death. After the drought of 1911–13 and the lifting of the beef tariff—both factors squeezing revenues—new president J. Leroy Nickel had tried to economize by cutting back operations, laying off a large number of employees, and cutting wages for those who remained. Morale suffered, and some of the company's most highly respected employees went elsewhere.

A crowning blow came with the death of the old patriarch in 1916. Litigation over federal inheritance taxes put a great financial burden on the firm through the 1920s, even as low postwar prices and rising expenses combined to shrink both income and asset values. Following the probate of Miller's will, multiple heirs pressed more than ten years of appeals. The outbreak of a serious epidemic of foot and mouth disease in California in 1924, as well as a new period of drought, forced the company to reorganize in 1925. Deeply in debt, Miller & Lux began to liquidate many assets in cattle and in land, but much of the cash ended in attorneys' pockets as the result of protracted litigations. In Oregon and Nevada the Pacific Live Stock Company gradually divested its holdings too. In 1933 Miller & Lux sold its breeding herd and "Double H" brand to Swift and Company.

The last of the Miller & Lux cattle in California were sold in 1938.

Nonetheless, Henry Miller's insistence on ironclad water rights continued to pay off, for those rights were in ever greater demand as the population of California grew, and the company's key position in the Central Valley was crucial to planning any large public water distribution system. In the early 1930s, after decades of investigation and discussion, the State of California proposed a massive system to transfer water from the wetter north to the drier south under the California Central Valley Project Act (1933). When the State was unable to sell bonds for the work, the U.S. Bureau of Reclamation assumed responsibility for the project in 1935. As part of the plan the federal government negotiated an agreement with Miller & Lux to buy water rights for $2.45 million. (It was not until 15 years later, after enormous changes engendered by World War II, that the Central Valley Project finally began water deliveries as a coordinated system. With the further development of statewide water deliveries made possible by the State Water Project of 1960, portions of former Miller & Lux land were eventually inundated beneath the San Luis Dam reservoir and associated canals.) Today some of the former Miller & Lux land remains in productive agriculture, but other land has been converted into the extensive San Luis National Wildlife Refuge Complex north of Los Banos, which includes some 45,000 acres of federally owned land along the San Joaquin River and 90,000 acres of private conservation easements, offering protection to huge numbers of migratory waterfowl and a variety of other wildlife, and rich recreational and educational opportunities for visitors.

After his death in 1916 Miller was laid to rest at the family compound on Mount Madonna, but when this property was sold in the 1920s the family removed his body to Woodlawn Memorial Park near Colma, where it rests today.[3] Miller's interment was one of the earliest there, and his pillared monument stands prominently near the entrance.

All three of Miller's homes vanished with the years. His Rincon Hill mansion was destroyed by the 1906 earthquake, and after decades of city renewal the site now lies beneath the ramps of the Bay Bridge. The big Bloomfield ranch house built in 1888 near Gilroy was also destroyed by fire in 1927 and never rebuilt, and Bloomfield Farm has passed into other hands. The site of the summer home at Mt. Madonna is now incorporated into a county park, marked only by a few concrete foundations in the forest.

Henry Miller, Jr.'s widow died in 1925, J. Leroy Nickel in 1937, and Nellie Miller Nickel in 1944. Continuing litigation by the next generation of heirs brought further division of the remaining Miller estate in the 1950s. The Miller & Lux company itself was formally dissolved in 1964, a little over a century after its founding. Large properties in the San Joaquin Valley went to separate branches of the family, the Nickels and the Bowles, who continue to be prominent in agriculture, land development, and philanthropy.[4]

Other remarkable characters in California's epic nineteenth-century agricultural expansion left legacies as well. California's premier pioneer John Bidwell managed his 22,000-acre Rancho Chico as a diversified farm with field and orchard crops, vineyards, and livestock, until his death in 1900 at the age of 80. Bidwell and his wife Annie made many contributions to the community of Chico, including the bequest of their home and Bidwell Park to the city.

In 1908 Claus Spreckels the "Sugar King" died, leaving a dynasty of descendants to carry on his name and enterprises. Some built their own reputations in society, politics, and philanthropy. His eldest son, John D., became celebrated as one of the developers of the city of San Diego, transforming it from a bankrupt village into a thriving commercial center. Rudolph was active as a businessman and civic reformer in San Francisco, while Adolph B. supported the development of Golden Gate Park and with his wife Alma financed the construction of the spectacularly sited Palace of the Legion of Honor, an art museum completed in 1924 on a bluff overlooking the ocean.

George Shima, the Delta potato farmer, became a highly respected leader in the Japanese community, his activities frequently reported in the press. In the 1920s Shima moved his family from their modest ranch house into a home in Berkeley, where he became one of the Bay Area's elder statesmen and patron of many good causes. Although he was unable to stem the tide of the era's anti-Japanese prejudice, he died in 1926 with an estate estimated at $15 million. The Shima Tract near Stockton still produces rich agricultural crops.

The Pan Pacific Exposition of 1915 celebrated a new era for California in more ways than one. Not only did the Panama Canal revolutionize inter-ocean shipping, it coincided with other transportation improvements. The railroads lost some of their monopoly over long-distance transportation as highway networks expanded, linking California in new ways with eastern markets. Shipping through San Francisco Bay continued to grow as steamship traffic changed the patterns of international trade; but the great wharves at Port Costa gradually fell silent and began to decay. Some of the old grain warehouses burned and were not rebuilt, and in the 1920s much of the Port Costa waterfront was pulled down. In 1927 Carquinez Strait Bridge, the first major bridge in the Bay Area, opened for traffic along the Lincoln Highway between Vallejo and Crockett. The railway bridge linking Benicia to the town of Martinez went into service in 1930, and the *Solano Ferry*'s last trip across the strait was on November 30 that year. In 1941 the original McNear dock and warehouse burned down, leaving little more than charred pilings dotting the shores west of Port Costa, mute reminders of a past era visible now only from the railroad tracks that curl along the shoreline.

Out in the rural areas of California the promise of irrigation began to transform agriculture. Where soils and water allowed, farmers converted and subdivided the former enormous wheat fields into more profitable plantings, and these crops found new markets in growing urban areas. Beginning in 1914 the University of California's Agricultural Extension Service greatly assisted farmers in making these conversions and taking advantage of the new knowledge accumulated through the Agricultural Experiment Station. After the "Great War" ended in 1918, Extension agents also helped farm families cope with the collapse of postwar agricultural demand.

San Francisco's Butchertown, the site for so many years of bawling animals and killing work, remained a slaughterhouse neighborhood for decades after the 1906 earthquake forced modernization, but its importance gradually declined until the area's 99-year lease expired in the 1960s. By then considered an area of "urban blight," the old neighborhood, now called the Bayview District, was slated for redevelopment. Few San Franciscans today can recall the days when cattle and cowboys moved through the streets.

Thus ends this journey into the past, when opportunity and character could converge to allow some men to build rural kingdoms out of wide open spaces and untested resources. Their successes helped create the underpinnings for the state's meteoric rise from an isolated raw frontier into a far broader economic world. Mineral wealth might soon be exploited, but the renewable gold of her agricultural industries would sustain California far longer. Early landowners like Glenn and Miller succeeded in accumulating vast tracts of land during the same years that allowed the robber barons of the east and the mining kings of the west to accrue vast fortunes. For a time their control over resources may have seemed to block others from pursuing their own dreams, yet both men acquired large portions of their holdings in relatively inhospitable areas where smaller farmers were unlikely to thrive, and neither engaged in land speculation. The Colusa plains and the southern San Joaquin Valley are still areas farmed largely by a limited number of farmers banking on economies of scale. Glenn and Miller worked hard for success under often difficult circumstances. Bolder than others, luckier perhaps too, Glenn and Miller built enterprises that in the long run contributed broadly to the California economy.

The story of Central Valley agriculture over the years is, of course, far more complex than the story told here. "Ít was the technological manipulation of water supplies than enabled the new West to blossom in the early decades of the 20th century," according to a study prepared for the U.S. Army Corps of Engineers as an introduction to its reprinting of the Alexander Report of 1873.[5] "California's agricultural development was not a saga of individual enterprise but the story of people who built organizations and institutions to control a problematic environment through use of technology." True: but individual enterprise as practiced in nineteenth-century America lay behind the decades of investigation, discussion, policy-making and compromise that ultimately transformed the Central Valley into a vast garden. Pioneers in the truest sense, Hugh Glenn and Henry Miller helped shape inland California at a pivotal time. Their own shadows on the land were the guarantee of their commitment, and their achievements proved what could be done. Others followed in their footsteps.

1 French, *Cattle County*, pp. 20–21 and 158–59.

2 See French; Gray.

3 Woodlawn was part of the new cemetery complex created at Colma for the relocation of San Francisco graveyards after 1914.

4 See Nickel, *Following the Cattle King.*

5 Jackson, Herbert, and Wee, *Engineers and Irrigation,* p. 3.

Bibliography

---------. "R. P. Effinger's Excellent Adventure: The Unknown Letters of a Young Ohio Lawyer, 1849–1850." *California History* 2004, Special issue, Vol. 82, No. 1.

Adler, Jacob. *Claus Spreckels, The Sugar King in Hawaii*. Honolulu: University of Hawaii Press, 1966.

Allen, Jimmy V. *Hugh Glenn*. M.A. thesis, Department of History, Sacramento State College, 1969.

Alexander, B.S., George H. Mendell, and George Davidson. *Report of the Board of Commissioners on the Irrigation of the San Joaquin, Tulare, and Sacramento Valleys of the State of California, 1873*. Annotated and with an introduction by W. Turrentine Jackson, Rand F. Herbert, and Stephen R. Wee. Engineer Historical Studies, Number 5: *Engineers and Irrigation*. Office of History, United States Army Corps of Engineers: Fort Belvoir, Virginia, 1990.

Ambrose, Stephen. E. *Nothing Like It in the World: The Men Who built the Transcontinental Railroad, 1863–69*. New York, Simon & Schuster, 2000.

Anderson, M. Kat. *Tending the Wild: Native American Knowledge and the Management of California's Natural Resources*. Berkeley: University of California Press, 2005.

Avella, Steven M. *Sacramento: Indomitable City*. Arcadia Publishing, 2003.

Baker, Mrs. Willis Kingsley. *Glenn History and Genealogy*. Presented to the Willows Library by the Willows chapter of the Daughters of the American Revolution, April 1977.

Bancroft, Hubert Howe. *History of California*. 7 volumes. San Francisco: History Company, 1884–1890.

Bancroft, Hubert Howe. *Chronicles of the Builders of the Commonwealth.* San Francisco: History Company, 1891–92. Vol. 3.

Beers, Terry, ed. *Gunfight at Mussel Slough: Evolution of the Western Myth.* Berkeley: Heyday Books, 2004.

Bowcock, Robert H. *Butchertown: A Collage of a San Francisco Institution during 1850–1969.* San Francisco: Falcon Books, 2004.

Brereton, Robert Maitland. "Proposal for English Upper Class Colonies in California." London, June 1875.

Bremer, Jeff. R. "The Trial of the Century...Lux v. Haggin," *Southern California Quarterly*, Summer 1999.

Brewer, William H. *Up and Down California in 1860–1864: The Journal of William H. Brewer.* Fourth edition, with maps. Ed. Francis P. Farquhar with a new foreword by William Bright. Berkeley: University of California Press, 2003.

Brimlow, George. *Harney County, Oregon, and Its Range Land.* Burns, Oregon: Harney County Historical Society, 1980.

Browning, Peter, ed. *San Francisco Yerba Buena: From the Beginning to the Gold Rush, 1769–1849.* Lafayette, CA: Great West Books, 1998.

Canright, Stephen. "Outpost of the Empire: British Dominance of the California Grain Trade." San Francisco Maritime National Historical Park Association *Sea Letter*, Number 63, November 2002.

Chan, Sucheng. *This Bittersweet Soil: The Chinese in California Agriculture, 1860–1910.* Berkeley: University of California Press, 1989.

Clay, John. *My Life on the Range.* Chicago: J. Clay, 1924.

Cleland, Robert Glass. *The Cattle on a Thousand Hills: Southern California, 1850–1870.* San Marino: The Huntington Library, 1941.

Cleland, Robert Glass. *The March of Industry.* Los Angeles: Powell Publishing Co., 1929.

Conmy, Peter Thomas. *The Beginnings of Oakland, California, A.U.C.* Oakland: Oakland Public Library, 1961.

Cook, Sherburne Friend. *The Population of the California Indians, 1769–1970.* Berkeley: University of California Press, 1976.

Cordray, William Woodrow. *Claus Spreckels of California.* Ph.D. dissertation, University of Southern California, 1955.

Cowan, Robert G. *Ranchos of California.* Los Angeles: Historical Society of Southern California, 1977.

Cronise, Titus Fey. *The Natural Wealth of California.* San Francisco: H.H. Bancroft, 1868.

Cronon, William. *Nature's Metropolis: Chicago and the Great West.* New York: W.W. Norton, 1991.

Dana, Richard Henry. *Two Years Before the Mast*. First published 1840, reprint New York: The Modern Library, 2001, with an introduction by Gary Kinder.

Davis, Horace. "Wheat in California." *Overland Monthly*, November 1868.

Davis, Horace. "Wheat in California: A Retrospect and Prospect." *Overland Monthly*, July 1898.

Davis, William Heath. *Seventy-Five Years in California, 1831–1906: A History of Events and Life in California: Personal, Political and Military; Under the Mexican Regime; During the Quasi-Military Government of the Territory by the United States, and after the admission of the State to the Union*. Edited by Douglas S. Watson. San Francisco: John Howell, 1929.

Deverell, William. *Railroad Crossing: Californians and the Railroad, 1850–1910*. Berkeley: University of California Press, 1994.

Dunn, Viola Jean. "Miller and Lux: A contribution toward a history." Manuscript, 1970. Bancroft Library.

Eigenheer, Richard Allen. *Early Perceptions of Agricultural Resources in the Central Valley of California*. Ph.D. dissertation, University of California, Davis, 1976.

Fraser, W.H. "Rainfall and Wheat in California." *Overland Monthly*, June 1899.

French, Giles. *Cattle Country of Peter French*. Portland, Oregon: Binfords & Mort, 1964.

Furer, Howard B. (Ed.) *Germans in America, 1607–1970: A Chronology and Fact Book*. Dobbs Ferry, NY: Oceana Publishers, 1973.

George, Henry. *Progress and Poverty*. San Francisco: W.M. Hilton, 1879.

Gilcrest, Evelyn. "Alvord Ranch Interlude: Life on a Celebrated Range." *Oregon Historical Quarterly*, December 1962.

Gillis, Michael J. and Magliara, Michael F. *John Bidwell and California: The Life and Writings of A Pioneer, 1841–1900*. Spokane: Arthur H. Clark Company, 2003.

Gray, Edward. *Life and Death of Oregon "Cattle King' Peter French, 1849–1897*. Eugene, Oregon: privately printed, 1995.

Gregor, Howard F. "The Industrial Farm as a Western Institution." *Journal of the West*, January 1970.

Green, Will Semple. *The History of Colusa County, California, and General History of the State*. San Francisco: Elliot and Moore, 1880. Reprint Sacramento, 1950.

Hall, Patricia A. *Jacinto: Book One, Mrs. H.J. Glenn*. Willows: privately printed, 2005.

Hanley, Michael F. (IV), *Tales of the I.O.N. Country*. Jordan Valley, Oregon: privately printed, 1988.

Haraszthy, Agoston. *Grape Culture, Wines and Wine Making.* New York: Harper & Bros, 1862.

Hastings, Lansford W. *The Emigrants' Guide to Oregon and California.* Reproduced in facsimile from the original edition of 1845, with historical notes and bibliography by Charles Henry Carey. Princeton: Princeton University Press, 1932.

Hayes, Derek. *Historical Atlas of California.* Berkeley: University of California Press, 2007.

Heizer, Robert F. & Albert B. Elsasser. *The Natural World of the California Indians.* California Natural History Guide #46. Berkeley: University of California Press, 1980.

Hilgard, E.W. et al, *Report on the Climatic and Agricultural Features and the Agricultural Practice and Needs of the Arid Regions of the Pacific Slope*, made under the direction of the Commissioner of Agriculture. Washington: Government Printing Office, 1882.

Hikert, Richard, and Oscar Lewis, eds. *Breadbasket of the World: California's Great Wheat-growing Era, 1860–1890.* San Francisco: Book Club of California, 1984.

Hittell, John S. *The Resources of California, comprising agriculture, mining, geography, climate, commerce, and the past and future development of the state.* San Francisco: A. Roman, 1863.

Hittell, John S. *The Commerce and Industries of the Pacific Coast of North America.* San Francisco, 1882.

Hoyt, Hugh Myron. *The Wheat Industry in California, 1850–1910.* M.A. thesis, Sacramento State College, 1953.

Hurtado, Albert L. *John Sutter: A Life on the North American Frontier.* Norman: University of Oklahoma Press, 2006.

Igler, David. *Industrial Cowboys: Miller & Lux and the Transformation of the Far West, 1850–1920.* Berkeley: University of California Press, 2001.

James, Dena. "Dr. Hugh J. Glenn: The Forgotten Legacy." Colusi County Historical Society, *Wagon Wheels*, 1984.

Johnson, Stephen, and Robert Dawson. *The Great Central Valley: California's Heartland.* A Photographic Project with Text by Gerald Haslam. University of California Press in association with the California Academy of Sciences, 1993.

Kelley, Robert L. *Battling the Inland Sea: American Political Culture, Public Policy, and the Sacramento Valley, 1850–1896.* Berkeley: University of California Press, 1989.

Kelley, Robert L. *Gold vs. Grain: The Hydraulic Mining Controversy in California's Sacramento Valley.* Glendale: A.H. Clark Co., 1959.

Kemble, John Haskell. *San Francisco Bay: A Pictorial Maritime History.*

Cambridge, Maryland: Cornell Maritime Press, 1957.

Kirk, Anthony. *A Flier in Oil: Adolph B Spreckels and the Rise of the California Petroleum Industry*. San Francisco: California Historical Society, 2000.

Latta, Frank. Collection in Huntington Library, San Marino, California.

Lawrence, William D. *Henry Miller and the San Joaquin Valley*. M.A. thesis, University of California, Berkeley, 1933.

Lotchin, Roger W. *San Francisco 1846–1856: From Hamlet to City*. New York: Oxford University Press, 1974.

Marks, B. "The Riparian Decision in Interior California," *Overland Monthly*, February 1887.

Marvin, Winthrop. *The American Merchant Marine, Its History and Romance from 1620 to 1902*. New York: Scribner's, 1902.

Matthews, Patrick. "Ferryboat *Solano*: The Largest Railroad Ferry in the World!" Seaways' *Ships in Scale*, January/February 2008.

McComish, Charles Davis, and Lambert, Rebecca. *History of Colusa and Glenn Counties, California*. Los Angeles: Historic Record Co., 1918.

McGie, Joseph F. *A History of Irrigation in Butte, Sutter, Glenn and Colusa Counties*. Gridley, CA: Joint Water Board District, 1980.

McGinty, Brian. *Strong Wine: The Life and Legend of Agoston Haraszthy*. Stanford University Press, 1998.

McGowan, Joseph A. *History of the Sacramento Valley*. 3 vol. New York: Lewis Historical Publishing Company, 1961.

Miller, Henry. Dictation (manuscript). Interview taken by Edwin Wilson Fowler for H.H. Bancroft's *Chronicles of the Builders of the Commonwealth*. Bancroft Library, University of California, Berkeley.

Miller, M. Catherine. *Flooding the Courtrooms: Law and Water in the Far West*. Lincoln, Nebraska: University of Nebraska Press, 1993.

Morgan, Dale L. *Jedediah Smith and the Opening of the West*. First published 1953, reprint by Bison Books, University of Nebraska Press, 1964.

Morgan, Dan. *Merchants of Grain*. New York: Viking Press, 1979.

Morison, Samuel Eliot. *The Oxford History of the American People*. New York: Oxford University Press, 1965.

National Park Service, U.S. Department of the Interior. "Grant-Kohrs Ranch." Informational pamphlet, Grant-Kohrs National Historic Site, Montana.

Nickel, George Wilmarth. *Following the Cattle King: A Lifetime of Agriculture, Water Management, and Water Conservation in California's Central Valley*. Berkeley: Regional Oral History Office, Bancroft Library, 2002.

Nordhoff, Charles. *California for Travellers and Settlers*. First published 1873, reprint Berkeley: Ten Speed Press, 1973.

Norris, Frank. *The Octopus: A Story of California*. New York: Doubleday & Co., 1901.

Orsi, Richard J. *Sunset Limited: The Southern Pacific Railroad and the Development of the American West, 1850–1930*. Berkeley: University of California Press, 2005.

Outcalt, John. *History of Merced County, California*. Los Angeles: Historic Record Co., 1925.

Parker, Paul Percy. "Management of Miller & Lux Ranches in California, Nevada, and Oregon: A Compilation of the letters of Henry Miller for use of the general manager and superintendent," Manuscript (1913?), California Historical Society.

Paul, Rodman. "The Great California Grain War," *Pacific Historical Review*, November 1958.

Paul, Rodman. "The Wheat Trade between California and the United Kingdom." *Mississippi Valley Historical Review*, December 1958.

Pisani, Donald J. *Water, Land, and Law in the West: The Limits of Public Policy, 1850–1920*. Lawrence, Kansas: University Press of Kansas, 1996.

Pisani, Donald J. *From Family Farm to Agribusiness: The Irrigation Crusade in California and the West, 1850–1931*. Berkeley: University of California Press, 1984.

Powers, Stephen. *Tribes of California*. First published by the U.S. Department of the Interior 1877, reprinted with an introduction and notes by Robert F. Heizer. Berkeley: University of California Press, 1976.

Pulling, Hazel Adele. *History of California's Range Cattle Industry*. Ph.D. dissertation, University of Southern California, 1944.

Rawls, James J. and Bean, Walton. *California: An Interpretive History*. Sixth Edition. New York: McGraw-Hill, Inc. 1993.

Rehart, Schyler and Patterson, William K. M. *Theo Kearney: Prince of Fresno*. Fresno City and County Historical Society, 1988.

Roberts, David. *A Newer World: Kit Carson, John C. Fremont, and the Claiming of the American West*. New York: Simon & Schuster, 2000.

Robinson, W.W. *Land in California: The Story of Mission Lands, Ranchos, Squatters, Mining Claims, Railroad Grants, Land Scrip, Homesteads*. Berkeley: University of California Press, 1948.

Rogers, Justus. H. *Colusa County: Its History and Resources*. Orland, California, 1891.

Rothstein, Morton. "The California Wheat Kings." #12 in Keepsake

Series, UC Davis Library Associates, 1987.

Royce, Josiah. *California: From the Conquest in 1846 to the Second Vigilance Committee in San Francisco.* Boston: Houghton Mifflin and Co., 1886.

Santos, Robert L. *The Eucalyptus of California: Seeds of Good or Seeds of Evil.* Denair, Calif: privately printed, 1997.

Sawyer, Charles, ed. *One Man Show: Henry Miller in the San Joaquin.* Los Banos: Milliken Museum Society, 2003.

Sawyer, Charles, ed. *The Road Over: Center Point–Old Los Banos–Pacheco Pass.* Los Banos: Milliken Museum Society, 2002.

Sawyer, Eugene T. *History of Santa Clara County, California.* Los Angeles: Historic Record Co., 1925.

Scheuring, Ann Foley. *Science and Service: A History of the Land-Grant University and Agriculture in California.* Berkeley: University of California ANR Publications, 1995.

Schmidt, Marjorie. "Through Other Eyes," *Fremontia*, April 2003.

Shumate, Albert. *Rincon Hill and South Park: San Francisco's Early Fashionable Neighborhood.* Sausalito: Windgate Press, 1988.

Simpson, Peter K. *The Community of Cattlemen: A Social History of the Cattle Industry in Southeastern Oregon, 1869–1912.* Moscow, Idaho: University of Idaho Press, 1987.

Smith, Kenneth Aubrey. *California: The Wheat Decades.* Ph.D. dissertation, University of Southern California, 1969.

Smith, Wallace. *Garden of the Sun: A History of the San Joaquin Valley, 1772–1939.* Second edition revised by William B. Secret, Jr. Fresno: Linden Publishing, 2004.

Starr, Kevin. *Inventing the Dream: California through the Progressive Era.* New York: Oxford University Press, 1985.

Stockbridge, Frank Parker. "Greatest Land Estate: Henry Miller, Cattle King, Had Vast Tracts, Which Are About To Be Sold Piecemeal." *The New York Times*, May 1, 1921.

Street, Richard Steven. *Beasts of the Field: A Narrative History of California Farmworkers, 1769–1913.* Stanford University Press, 2004.

Street, Richard Steven. "The Murder of Hugh Glenn." *Sacramento Magazine*, July 1985.

Taper, Bernard. "The King of Ranchers." *American Heritage Magazine*, August 1967.

Taylor, Paul. "Foundations of California Rural Society." *California Historical Society Quarterly*, Fall 1945.

Thompson, Kenneth. "Historical Flooding in the Sacramento Valley," *Pacific Historical Review*, September 1960.

Torchiana, H.A. van Coenen. *California Gringos.* San Francisco: P. Elder and Co., 1930.

Treadwell, Edward F. *The Cattle King: A Dramatized Biography.* First published 1931, fifth printing San Francisco: Western Tanager Press, 1981.

Vaught, David. *After the Gold Rush: Tarnished Dreams in the Sacramento Valley.* Baltimore: The Johns Hopkins University Press, 2007.

Warren, John Quincy Adams. *California Ranchos and Farms, 1846–1862.* Madison: State Historical Society of Wisconsin, 1967. Edited with an introduction by Paul W. Gates.

Wells, Ronald A. "Revising the Sesquicentennial Narrative: The Importance of Josiah Royce's *California* for Our Time." *California History*, Summer/Fall 2001.

White, Thelma Bulkley. *Glenn County Sketchbook.* Ed. Dorothy H. Hill. Chico, CA: The Branch, 1995.

Wik, Reynold. M. "Some Interpretations of the Mechanization of Agriculture in the Far West," *Agricultural History,* 1975.

Wooldridge, Jesse Walton. *History of the Sacramento Valley.* Biographical, Vol. 2. Chicago: Pioneer Historical Publishing Co., 1931.

Yoshimura, Toshio. *George Shima, Potato King and Lover of Chinese Classics.* English translation from Japanese original. Fukushima, Japan: 1981. [Copy in Huntington Library]

ANN FOLEY SCHEURING was born in southeastern Oregon and educated in Portland. After a Fulbright year in Germany, she earned graduate degrees at both UC Berkeley and UC Davis, where she was a senior editor and writer for many years. She has produced several books including *A Guidebook to C alifornia Agriculture*; *Science and Service*; and *Abundant Harvest: The History of the University of California, Davis*. Her longstanding interest in history and agriculture has roots in her family's commitment to farming since the 1890s. She lives on a farm in the Capay Valley with her husband David.

Index